Man's place in evolution

British Museum (Natural History)
Cambridge University Press

Published by the British Museum (Natural
History), London and the Press Syndicate of the
University of Cambridge
The Pitt Building, Trumpington Street,
Cambridge CB2 1RP
32 East 57th Street, New York, NY 10022, USA
296 Beaconsfield Parade, Middle Park, Melbourne
3206, Australia

First published 1980

**British Library Cataloguing
in Publication Data**

Man's place in evolution
1. Human evolution
I. British Museum (Natural History)
573.2   GN281   80-41067
British Museum (Natural History)
Man's Place in Evolution
Includes index.

ISBN 0 521 23177 9 hard covers
ISBN 0 521 29849 0 paperback

Printed in Great Britain by Balding & Mansell Ltd., Wisbech, Cambs.

Contents

Preface

Man – *Homo sapiens* – is only one of many thousands of animal species alive today. How are we human beings related to other living animals? And how are we related to the various 'fossil men' whose remains have been found? This book will help you to decide for yourself.

It begins by explaining how animals can be grouped together on the basis of the characteristics that they share, and shows that all human beings are vertebrates ... mammals ... primates ... and apes. It then describes a simple method for working out the relationship between different animals, and applies this method to the apes in an attempt to discover which are our closest living relatives.

The book then turns to 'fossil man', but it makes no attempt to reconstruct the history of human evolution. Instead, it looks more closely at man's unique characteristics, and then examines the fossil remains for evidence of these characteristics. In this way a picture is built up of how modern human beings might be related to the ancient ramapithecines and australopithecines, and to extinct human beings such as the habilines and the neandertals. There are many diagrams, colourful reconstructions and photographs – including photographs of classic fossil specimens from all over the world.

This is a companion book to the exhibition **Man's place in evolution**, which opened at the Natural History Museum in May 1980. Like the exhibition, it was planned with the guidance of Museum experts, particularly from the Museum's Sub-Department of Anthropology. A great many people, both within the Museum and outside, have helped in the preparation of **Man's place in evolution**, and I should like to take this opportunity of thanking everyone concerned.

R. H. HEDLEY Director
British Museum (Natural History)
June 1980

If life on Earth appeared only once, then all living things must be descended from a single common ancestor, and all living things – including human beings – must be related to each other.

In this book we shall try to discover:

Which animals are our closest living relatives?

Do we have any closer relatives amongst the fossils?

Chapter 1
Man is an animal

9

Human beings ('man') can be studied in much the same way as any other animals. A biologist, for example, might describe modern man like this:

- Adult height ranges from 1.2 metres (pygmy race) to 2.0 metres (tallest African race)

- Upright body position

- Bipedal method of locomotion ('walking')

- Skin colour variable, from very pale to black

- Body hair sparse except under arms and in pubic region. Head hair long (mane), with beard and moustache in adult males

- Mixed diet, including fruit, vegetables and meat (usually cooked)

- Distribution worldwide

One species

All living human beings belong to one species – *Homo sapiens*. The biological definition of an animal species states that 'all members of the species can interbreed'. This is true for human beings. There is no biological reason why a man and a woman from different races should not interbreed.

What sort of animal?

There are countless different kinds of animals alive today, of many different shapes and sizes. In order to make sense of this bewildering variety, we **classify** animals. We sort them into groups on the basis of characteristics that they share with one another.

How do we classify ourselves? Which group of animals does man belong to?

Man is a vertebrate

Broadly speaking, animals can be divided into two groups: those that have backbones, and those that do not. Animals that have backbones are called **vertebrates**.

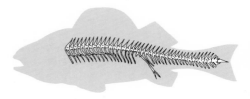

Human beings have backbones, so we are vertebrates too. We share this characteristic with all other vertebrates.

Backbones are so complicated, and so similar in all the animals that have them, that we believe they have evolved only once. So we can consider that all vertebrates are descended from a common ancestor that had a backbone.

Vertebrates are therefore more closely related to each other than to any other animals that do not have a backbone.

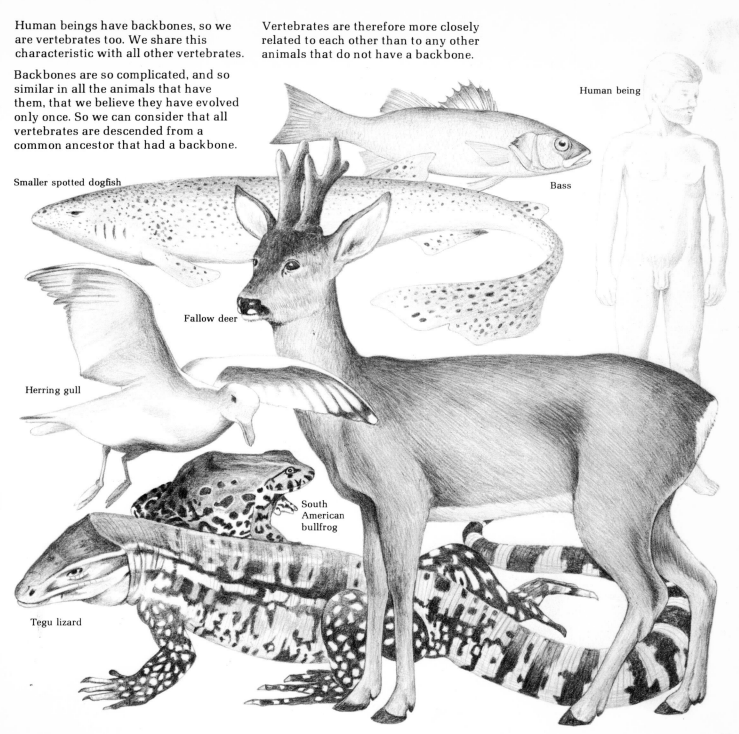

Human being

Smaller spotted dogfish

Bass

Fallow deer

Herring gull

South American bullfrog

Tegu lizard

Man is a mammal

We can sort the vertebrates into several different groups on the basis of other characteristics that they share with one another.

For example, **mammals** are vertebrates with . . .

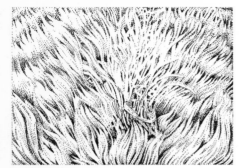

• fur or hair

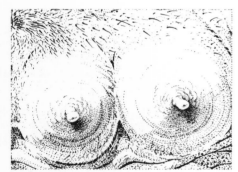

• milk-producing glands

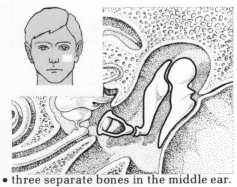

• three separate bones in the middle ear.

Human beings have all these characteristics, so we are mammals too. Because mammals share all these characteristics, we think that they are more closely related to each other than they are to any other animals. So we are more closely related to dogs and cats, for example, than we are to fishes or frogs.

Diana monkey

Koala bear

Spiny anteater or Echidna

Malaysian fruit bat

Human being

Hedgehog

Leopard

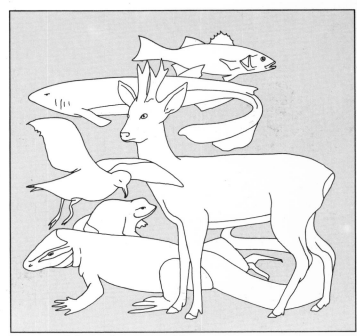

Can you pick out the mammal from this group of vertebrates? The answer is on p. 106.

Man is a primate

We can sort the mammals themselves into different groups by putting together all the ones that share certain characteristics.

For example,
primates are mammals that have...

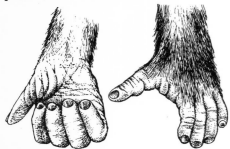

- fingernails and toenails, instead of claws

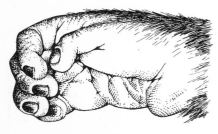

- an opposable thumb... or an opposable big toe (an opposable thumb is one that can touch the fingers of the same hand)

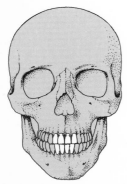

- four incisors (biting teeth) in both the upper and the lower jaw.

What about human beings?
Are we primates?
Check for yourself.
Do you have fingernails?
Can you make your thumb touch the fingers of the same hand?
How many incisors do you have in each jaw?

Human beings have all these primate characteristics, so we are primates too. Because primates share all these characteristics, we think that they are more closely related to each other than they are to any other animals.

White-handed gibbon

Red howler monkey

Tarsier

Edward's potto

14

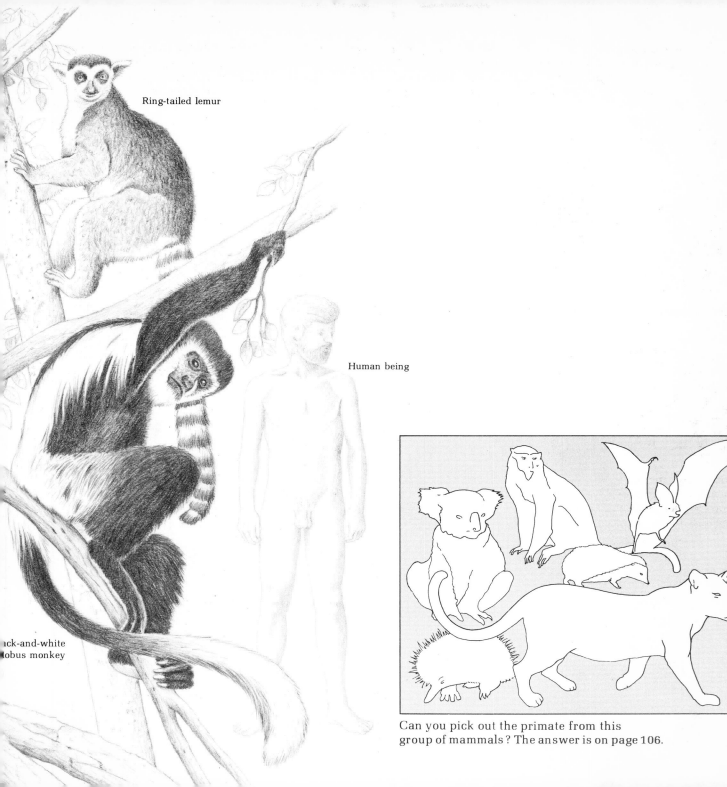

Ring-tailed lemur

Human being

ack-and-white
obus monkey

Can you pick out the primate from this
group of mammals? The answer is on page 106.

Man is an ape

If we look more closely at the primates, we can see that they too can be sorted into a number of different groups. One of these groups is the **apes.**

Apes are primates that have...

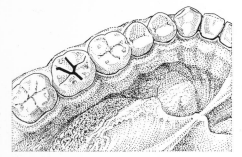

- a Y-shaped pattern on the surface of their molars (chewing teeth)

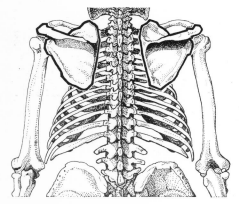

- shoulder blades at the back, not at the sides

- no tail

White-handed gibbon

Human being

Chimpanzee

Gorilla

But remember,
an ape is a primate
and a primate is a mammal
and a mammal is a vertebrate

So an ape has all these
characteristics:

- a backbone

Orang-utan

- fur or hair

- milk-producing glands

- 3 separate bones in the middle ear

- fingernails and toenails

- an opposable thumb ... or big toe

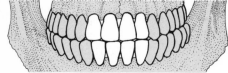

- 4 incisors in the upper and lower jaw

Look closely at these primates.
Can you pick out the ape?
The answer is on page 106.
What about you? Are human beings apes?

The apes

The apes form quite a small group of
primates: chimpanzees, gibbons,
gorillas, human beings ('man'), and
orang-utans. The apes share
characteristics (such as the Y-shaped
pattern on the molars) that no other
animals share. So we think that they
are more closely related to each other
than they are to any other animals.

Can we find out how the apes are
related to each other? In particular,
can we find out which of the other
apes is most closely related to man?

We cannot go back in time and see how animals evolved, so we can never be sure how the apes – or any other animal species – are related to each other. But we can suggest possible relationships, and then test them to find out which one is the most likely.

How do we begin?

When we are trying to work out the relationships between different species, it is easier if we begin by making two assumptions.

First, we assume that new species arise when one species splits into two, like this

This assumption allows us to **test** the relationships we suggest, because it means that every species must have a 'closest relative'.

Second, we assume that none of the species we are considering is the ancestor of any of the others. This is a fairly safe assumption, because so many animals have lived and died during the history of life on Earth that the chances of finding – and recognizing – any particular ancestral species are very, very small.

So, if we are considering two species

we assume that they are related like this

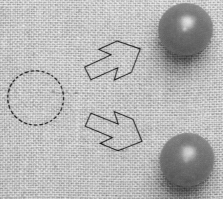

(The broken circle represents their ancestor.)

and **not** like this

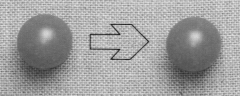

And if we are considering **three** species

we assume that they are related like this

(The broken circles represent the ancestors.)

In other words, we assume that two of the species share a common ancestor that is not shared by the third species.

This relationship can be represented by a simple branching diagram called a **cladogram**.

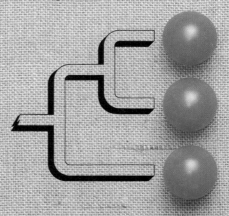

Possible relationships

Let's call our three species A, B and C. They could be related to each other in any one of the following three ways.

1 A and B could be more closely related to each other than either of them is to C.

This relationship can be shown like this

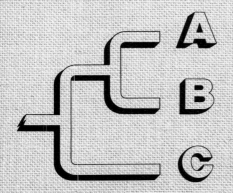

or like this

2 A and C could be more closely related to each other than either of them is to B.

This can be shown like this

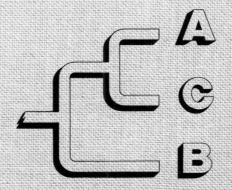

or like this

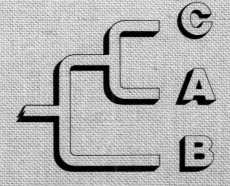

3 B and C could be more closely related to each other than either of them is to A.

This can be shown like this

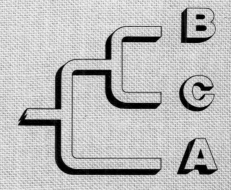

or like this

We should test each of these possible relationships, **1**, **2** and **3**, to see which one is most likely.

How do we test relationships?

We test relationships between species by looking for characteristics that are similar in different species because they have been inherited from a common ancestor. These are called **homologies**.

The more homologies that two species share, the more closely they are related. And two species that are 'closest relatives' share homologies that are not shared by any other species. In other words, they share **unique homologies**.

Looking for homologies

We can look for homologies in a wide variety of characteristics, ranging from molecules to whole organs.

1 Bones and teeth

Characteristics of bones and teeth can provide good examples of homologies. One of the best examples is the backbone that all vertebrate animals share.

Homologies in bones and teeth are very important when we are considering fossil animals, because bones and teeth are often the only parts of the animal that are preserved by fossilization.

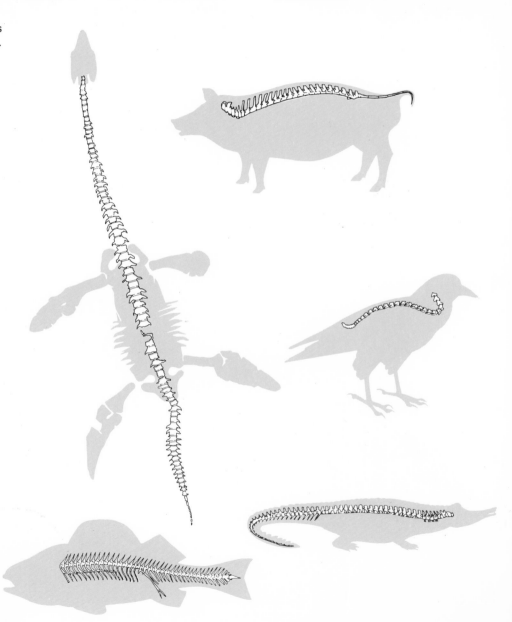

Another homology, shared by some of the apes, is the **frontal sinus** – a cavity in the skull, just above the eye. Human beings, chimpanzees and gorillas all share this homology, but gibbons, orang-utans and most other primates do not have a true frontal sinus.

But we must be careful. Not all similarities of bones and teeth are homologies. Some similarities may have evolved independently. They may be adaptations to a similar way of life – the wings of bats, birds and pterodactyls, for example. Such similarities cannot be used to test relationships.

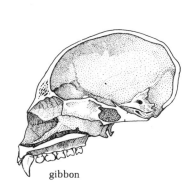

gibbon

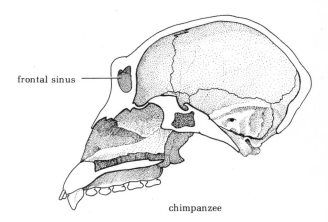

frontal sinus

chimpanzee

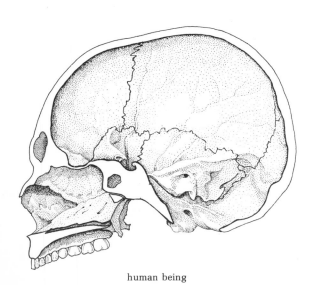

human being

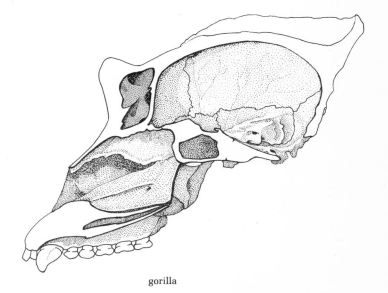

gorilla

2 Molecules

We can find examples of homologies in molecules. For example, apes all have a particular kind of blood protein molecule (delta chain haemoglobin) that the other mammals do not have.

Another example involves the alpha chain of the haemoglobin molecule. This chain is made up of 141 amino acids. Human beings and chimpanzees have identical alpha chains. But, in other primates, some of the amino acids are different, or are in a different order along the chain.

When two or more species share the ability to make a complex molecule like this, it is almost certain that they have inherited this ability from their common ancestor. Making molecules is such a complicated process that it is very unlikely that the same complex molecule would evolve independently in different species.

3 Soft parts of the body

Homologies can also be found in the soft parts of the body. The milk-producing glands of mammals are a good example.

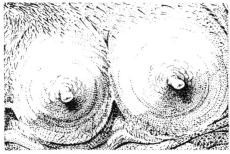

milk-producing glands

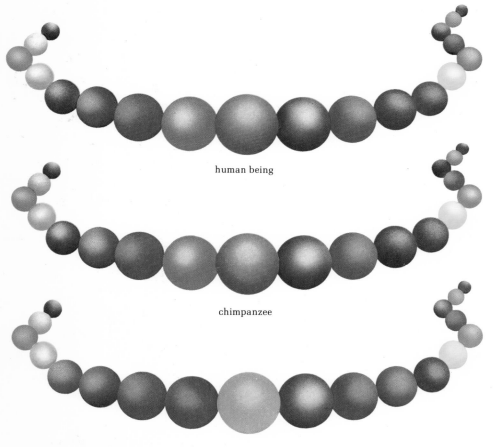

human being

chimpanzee

gelada baboon

These diagrams represent the sequence of amino acids in just one part of the alpha chain.

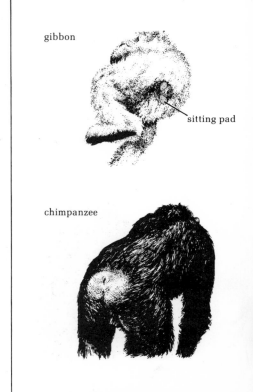

gibbon

sitting pad

chimpanzee

Another example is the **sitting pads** of hard skin ('ischial callosities') that most monkeys have on their rumps. Gibbons also share this homology, but the other apes do not.

But of course not all similarities between the soft parts of different animals are homologies. The parts may be similar because they have a similar function and *not* because they are inherited from a common ancestor. An example is the webs between the toes of ducks, frogs and seals.

4 Chromosomes

We can also find homologies in chromosomes. Similarities between the chromosomes of different species are probably inherited from a common ancestor. It is very unlikely that the similarities have evolved independently in each species, and the chromosomes look like each other just by chance.

An example is the total number of chromosomes that an animal has in each of its cells. Human beings have 46 chromosomes (23 pairs) in each body cell. Chimpanzees and gorillas have 48 (24 pairs).

macaque monkey

gorilla

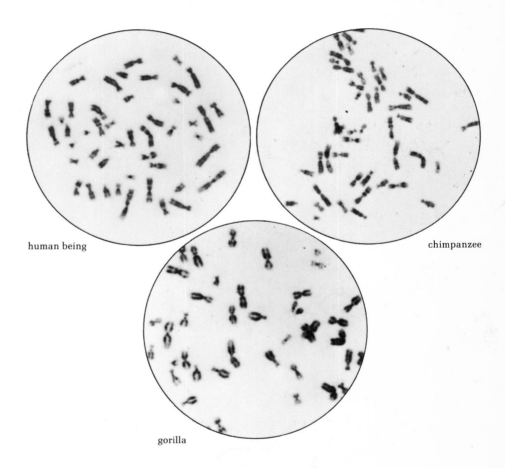

human being

chimpanzee

gorilla

We can use these different kinds of homologies to test our ideas about the relationships between different species. The more homologies we find, the more confident we are that our ideas are correct.

Chapter 3
Man's closest
living relatives

Which of the other apes are our closest relatives? ... the gibbons, the orang-utans, the gorillas or the chimpanzees?

To answer this question, we must find out if any of the other apes share unique homologies with human beings. If any of the other apes do share unique homologies with us, then they must be our closest relatives (see page 22).

Our close relatives

Evidence suggests that gibbons and orang-utans are not as closely related to us as chimpanzees and gorillas are.

For example, human beings, chimpanzees and gorillas all have a frontal sinus (see page 23). Gibbons, orang-utans and other primates do not have a true frontal sinus. This is evidence that chimpanzees and gorillas are more closely related to us than gibbons and orang-utans are.

And evidence from tests on certain blood plasma proteins (fibrino-peptides) suggests that orang-utans are more closely related to us than gibbons are.

We can represent these relationships by a cladogram like this ...

human beings, gorillas, chimpanzees

orang-utans

gibbons

White-handed gibbon
Hylobates lar

Orang-utan
Pongo pygmaeus

28

Our closest living relatives

Chimpanzees and gorillas are more closely related to us than any other living animals are. But which of them are our **closest** living relatives?

Chimpanzee
Pan troglodytes

Gorilla
Gorilla gorilla

Is it the gorillas?

or the chimpanzees?

or both gorillas **and** chimpanzees?

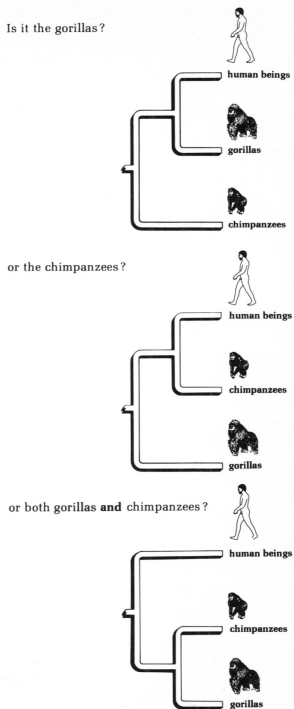

human beings

gorillas

chimpanzees

human beings

chimpanzees

gorillas

human beings

chimpanzees

gorillas

29

To try to answer this question, we must look carefully at all three animals for evidence of shared characteristics.

The traditional view is that chimpanzees and gorillas are more closely related to each other than either of them is to human beings (cladogram 3). This view is based on similarities between their bones, teeth and soft parts of the body.

But, recent evidence from studies on chromosomes and molecules has begun to challenge the traditional view.

In assessing the evidence, we must consider whether or not the shared characteristics that we are considering really are homologies, and not merely adaptations to a similar way of life.

And to make sure that we are considering only unique homologies, we must check to see whether or not they are shared by other primates too. If a characteristic is shared by other primates, it is obviously not unique. So it cannot be used to sort out the relationships between human beings, chimpanzees and gorillas.

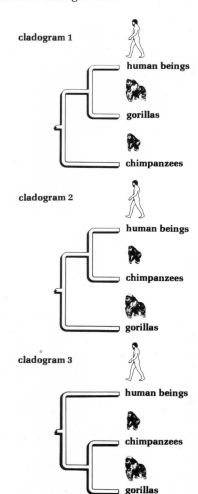

cladogram 1

human beings

gorillas

chimpanzees

cladogram 2

human beings

chimpanzees

gorillas

cladogram 3

human beings

chimpanzees

gorillas

The table on the opposite page includes examples of different kinds of shared characteristics. Study it carefully and see if you can decide for yourself which are our closest relatives.

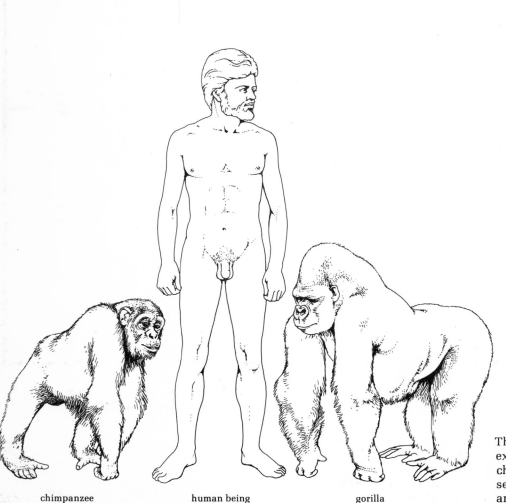

chimpanzee   human being   gorilla

Shared characteristics	Other primates	Gorillas	Human beings	Chimpanzees	According to this evidence, which cladogram is correct?
Bones and teeth					
limb length	arms and legs equal	legs shorter than arms	arms shorter than legs	legs shorter than arms	**3**
canine teeth	large	large	small	large	could be **1, 2** or **3**
thumbs	long	short	long	short	**3**
Soft parts of the body					
head hair	short	short	long	short	could be **1, 2** or **3**
calf muscles	small	small	large	small	could be **1, 2** or **3**
buttocks	thin	thin	fat	thin	could be **1, 2** or **3**
Chromosomes					
total number	42 or more	48	46	48	**3**
structure of chromosomes 5 and 12		different from other primates	like other primates	different from other primates	**3**
'fluorescence' of chromosomes Y and 13		same as human beings	same as gorillas	like other primates	**1**
Molecules					
alpha-haemoglobin chain, compared with that of a human being	several differences	one amino acid different		identical	**2**
'GM factor' in blood	not variable	not variable	same variability as chimpanzees	same variability as human beings	**2**
sequence of amino acids in myoglobin		like chimpanzees	generally like other primates		**3**

On the basis of the evidence in the table, which do you think are our closest living relatives? Chimpanzees, gorillas, or both?

Obviously it is very difficult to decide and, whatever you have decided, you could well be right. And even if the table had included all the evidence available, it would still have been difficult to decide, because much of the evidence is conflicting, and the picture is not yet complete.

At present, we are not sure which are our closest relatives. But, for the sake of argument, throughout the rest of this book we will assume that our closest living relatives are **both chimpanzees and gorillas.**

The next question is
Do we have any closer relatives amongst the fossils?

The fossilized remains of many different kinds of extinct apes have been found in Africa, Europe and Asia. Could any of these extinct apes be more closely related to human beings than chimpanzees and gorillas are?

To find out, we must see if they share any of the characteristics of human beings that chimpanzees and gorillas do not share.

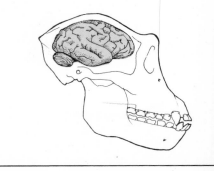

34

Unlike chimpanzees and gorillas, human beings

 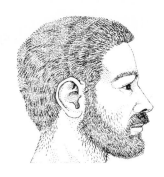

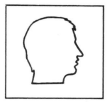

● have very short muzzles — flat faces,

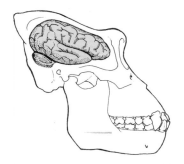 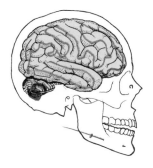

● they walk upright on two legs,

 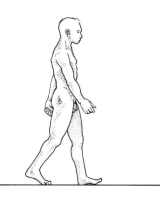

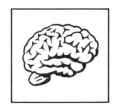

● and they have large brains.

If a fossil ape shows evidence of any of these characteristics, we can be fairly certain that it is more closely related to us than chimpanzees and gorillas are.

The ramapithecines

The ramapithecines are extinct apes that lived between about 14 million and 8 million years ago. Their fossil remains – mostly jaw fragments – have been found in many parts of the world, especially in Africa, the Near East and Asia.

The name **ramapithecine** comes from *Rama* – an Indian god, and *pithekos* – a Greek word meaning monkey or ape.

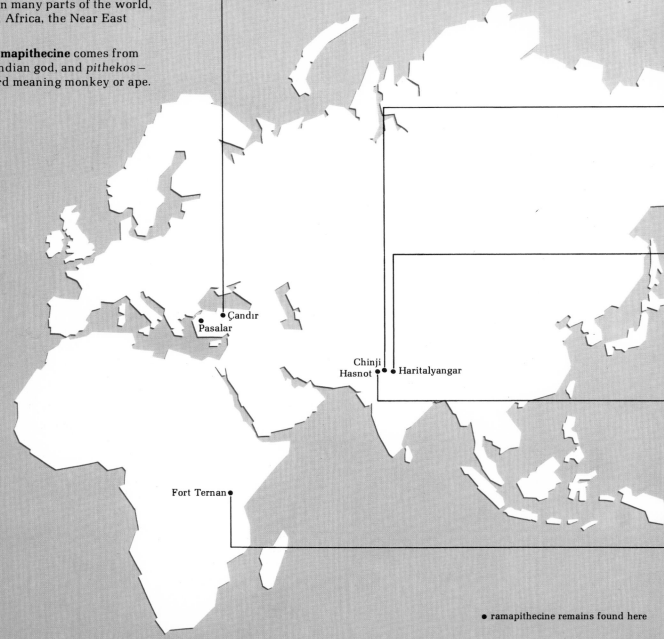

Çandır
Pasalar

Chinji
Hasnot ● ● Haritalyangar

Fort Ternan ●

● ramapithecine remains found here

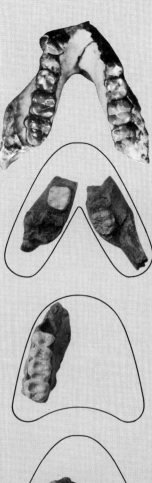

Lower jaw of a ramapithecine from Çandır, Turkey. One of the most complete ramapithecines found so far.

Fragments of the lower jaw of a ramapithecine from Chinji, Pakistan.

Fragment of the upper jaw of a ramapithecine from Haritalyangar, India.

Part of the lower jaw of a ramapithecine from Hasnot, Pakistan.

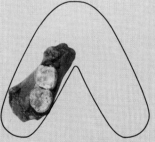

Fragments of the upper jaw of a ramapithecine from Fort Ternan, Kenya.

How are the ramapithecines related to man?

Let's suppose that the ramapithecines are more closely related to us than chimpanzees and gorillas are.

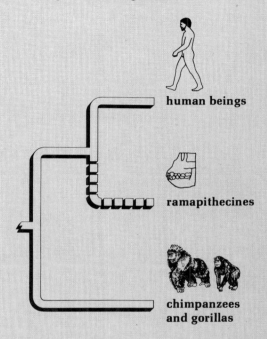

human beings

ramapithecines

chimpanzees and gorillas

To test this idea, we must find out if they share any of the characteristics of human beings that are not shared by either chimpanzees or gorillas.

Most of the ramapithecine remains that have been found so far consist only of fragments of jaws. But even these small fragments can give us clues. They can tell us for example whether the ramapithecines had short muzzles like us, or long muzzles, like chimpanzees and gorillas.

Different muzzles . . . different teeth

Human beings have short muzzles – flat faces.

And, associated with their flat faces, human jaws and teeth have several characteristics that chimpanzees and gorillas do not have.
These characteristics are linked to the way human beings chew:

When a chimpanzee chews:

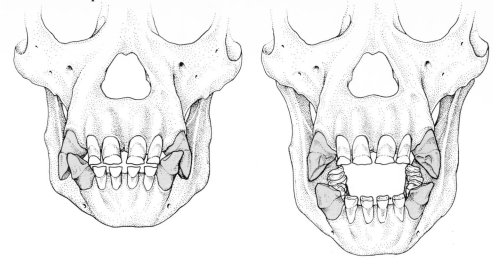

- its large canine teeth overlap, preventing side-to-side movement

- so its jaws move up and down

When a human being chews:

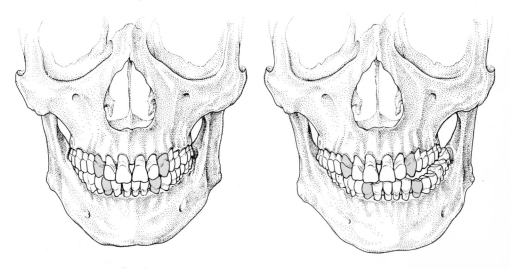

- the canine teeth do not overlap

- so the jaws can move from side to side

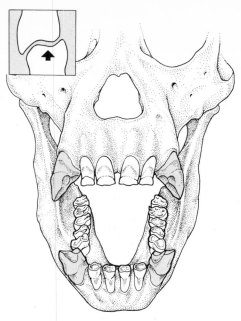

- the chewing teeth all meet at the same time, because they are in two straight rows

- the chewing teeth wear down unevenly

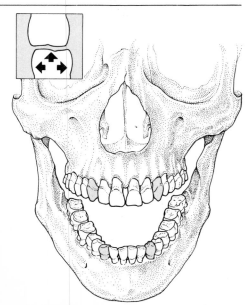

- the chewing teeth wear down flat

You can see evidence for these different ways of chewing if you look at the jaws and teeth ...

chimpanzee jaw ($\frac{1}{3}$ actual size)

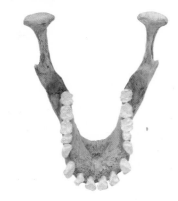

Notice that

- the chewing teeth are worn down unevenly

- the canine teeth are large

- the chewing teeth are in two straight rows

human jaw ($\frac{1}{3}$ actual size)

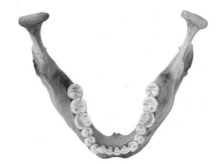

Notice that

- the chewing teeth are worn down flat

- the canine teeth are small

- the teeth form a rounded arc

Did the ramapithecines have teeth like us?

Look closely at this ramapithecine jaw, especially at the chewing teeth. Compare it with the human and chimpanzee jaws.

The ramapithecine chewing teeth are worn down flat. It probably chewed from side to side like a human being.

ramapithecine jaw (actual size)

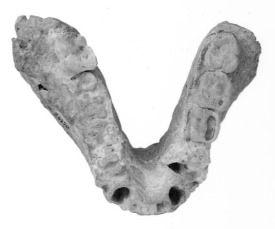

Ramapithecine jaw
from Potwar, in Pakistan

Did the ramapithecines have short muzzles?

A complete ramapithecine muzzle has not yet been found. But scientists have reconstructed one from fragments of jaws...

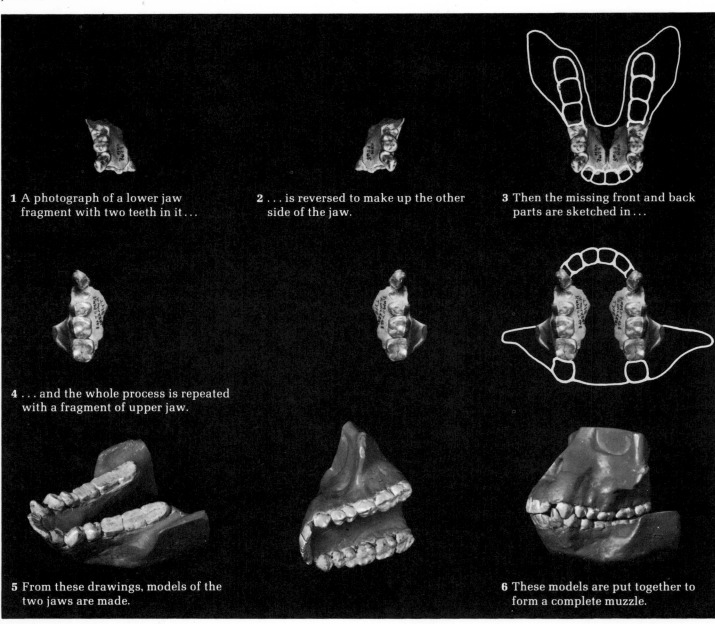

1 A photograph of a lower jaw fragment with two teeth in it...

2 ...is reversed to make up the other side of the jaw.

3 Then the missing front and back parts are sketched in...

4 ...and the whole process is repeated with a fragment of upper jaw.

5 From these drawings, models of the two jaws are made.

6 These models are put together to form a complete muzzle.

Compare the reconstructed rama-pithecine muzzle with the muzzles of a human being and a chimpanzee. To help you, all three skulls have been drawn the same size.

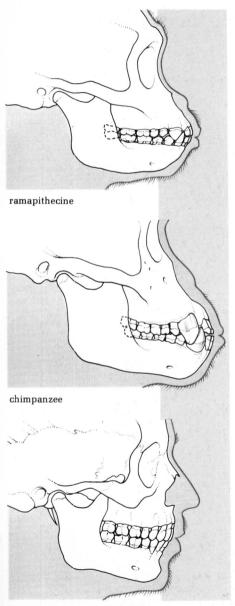

ramapithecine

chimpanzee

human being

The ramapithecine muzzle is fairly short – more like the muzzle of a human being than the muzzle of a chimpanzee or a gorilla.

All the evidence so far suggests that the ramapithecines are more closely related to human beings than chimpanzees and gorillas are . . .

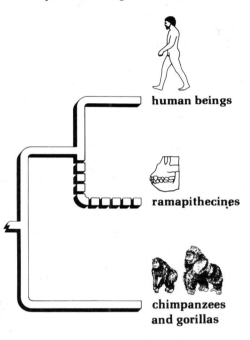

human beings

ramapithecines

chimpanzees and gorillas

Of course, it is always possible that the ramapithecine's short muzzle is not a homology at all. It may be pure chance that the ramapithecines have short muzzles like us.

Another possibility is that the common ancestor of the ramapithecines and other apes had a short muzzle. In other words, a short muzzle may be a primitive characteristic, and the long muzzles of chimpanzees and gorillas may have evolved more recently.

In either case, chimpanzees and gorillas **could** be more closely related to us than the ramapithecines are . . .

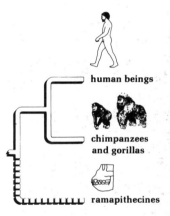

human beings

chimpanzees and gorillas

ramapithecines

To test this idea, we need more fossil evidence than we have at present. So, in the rest of the book, we will assume that the ramapithecines are more closely related to us than chimpanzees and gorillas are.

41

The australopithecines

The australopithecines lived between about 5 million and 1.5 million years ago. Their fossil remains have been found in many parts of Africa, especially along the Rift Valley.

The name **australopithecine** comes from *austral*, which is Latin for southern, and *pithekos*, which is Greek for ape.

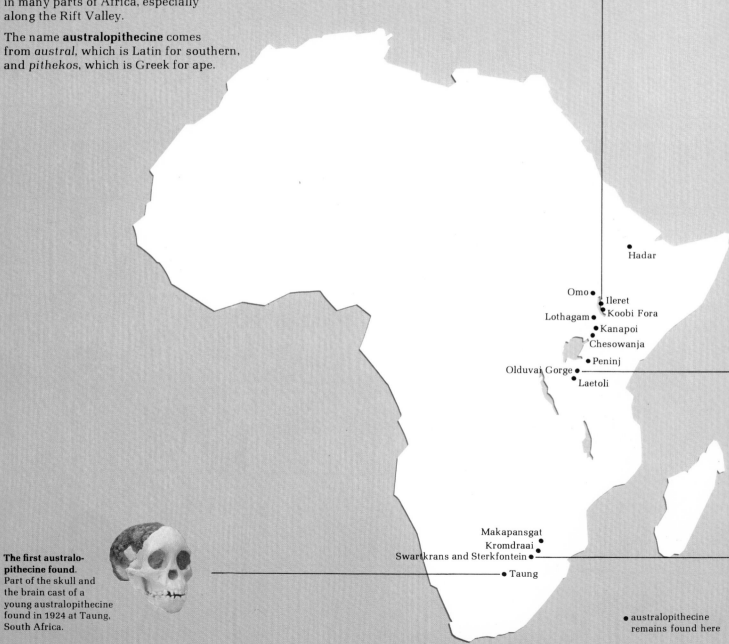

Hadar

Omo
Ileret
Lothagam
Koobi Fora
Kanapoi
Chesowanja
Peninj
Olduvai Gorge
Laetoli

Makapansgat
Kromdraai
Swartkrans and Sterkfontein
Taung

● australopithecine
remains found here

The first australopithecine found. Part of the skull and the brain cast of a young australopithecine found in 1924 at Taung, South Africa.

42

Part of the skull of an
australopithecine from
Ileret, East Turkana,
Kenya.

Part of the skull of a
robust australopithecine
from Olduvai Gorge,
Tanzania.

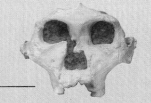

Part of the skull of a
robust australopithecine
from Swartkrans,
South Africa.

How are the australopithecines related to man?

Let's suppose that the australopithecines are more closely related to
human beings and ramapithecines
than chimpanzees and gorillas are.

To test this idea, we must find out if
the australopithecines share any of the
characteristics of human beings and
ramapithecines that are not shared by
chimpanzees or gorillas.

Did they have short muzzles?

This model of a female australopithecine has been reconstructed
using measurements taken from
australopithecine bone fragments
found at Sterkfontein in South Africa.
Study her carefully. You can see that
she has a short muzzle, like a human
being or a ramapithecine – not a long
one like a gorilla or a chimpanzee.

So the australopithecines probably **are**
more closely related to human beings
and ramapithecines than chimpanzees
and gorillas are.

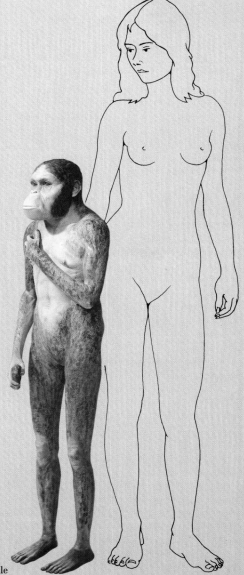

Adult female
australopithecine

Modern human being

But which are **more** closely related to
human beings?

the ramapithecines?

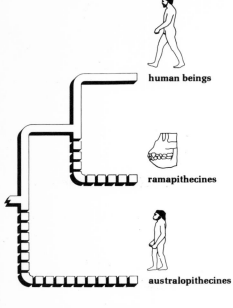

human beings

ramapithecines

australopithecines

or neither?

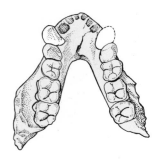

human beings

australopithecines

ramapithecines

or the australopithecines?

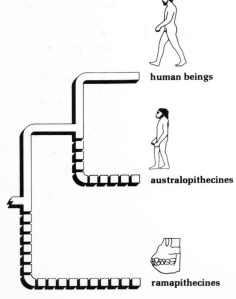

human beings

australopithecines

ramapithecines

Look carefully at these ramapithecine
and australopithecine jaws – especially
at the cusp pattern on the surface of
the premolar teeth. Which is more like
the human jaw?

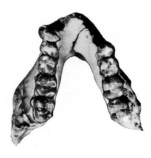

Ramapithecine jaw
from Çandır, Turkey

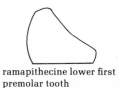

ramapithecine lower first
premolar tooth

To answer this question, we must
decide which of them shares more of
man's special characteristics. Almost
the only ramapithecine remains we
have found so far are jaws and teeth.
We must compare these with the jaws
of australopithecines to see which is
more like a human jaw.

Both the australopithecine jaw and the human jaw have two cusps on each premolar tooth. But the ramapithecine lower first premolars have only one cusp.

This supports the idea that the australopithecines are more closely related to us than the ramapithecines are ...

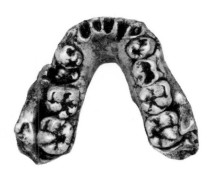

Jaw of a young australopithecine from Makapansgat, South Africa

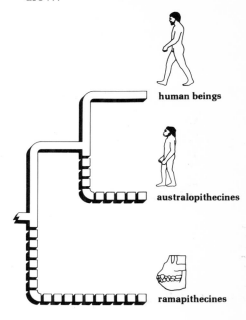

Human jaw

human beings

australopithecines

ramapithecines

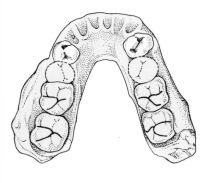

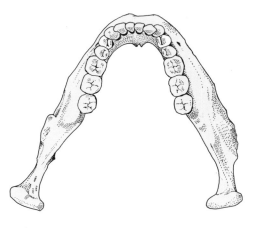

Is there any other evidence to suggest that the australopithecines are closely related to human beings? Fortunately, fragments of most parts of the australopithecine skeleton have been found. So we can look for evidence of man's other important characteristics. For example, we can look for evidence that they shared our characteristic way of walking – upright on two legs.

australopithecine lower first premolar tooth

human lower first premolar tooth

45

How did the australopithecines walk?

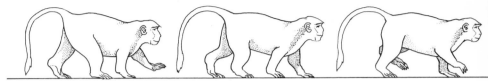

Did they walk on all fours
like monkeys?

Or did they knuckle-walk
like gorillas and chimpanzees?

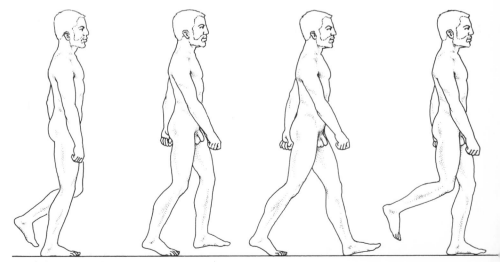

Or did they walk upright
like human beings?

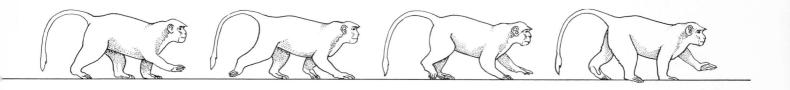

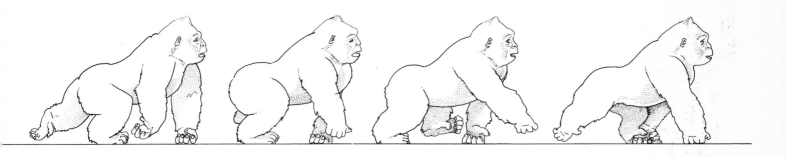

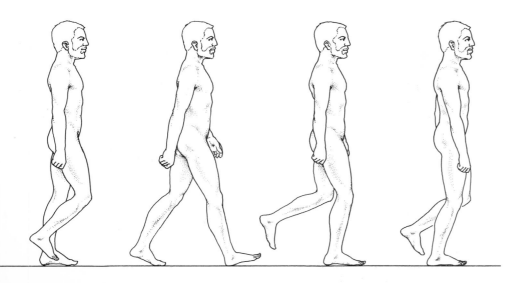

The earliest apes probably walked on all fours. But human beings and gorillas have each evolved their own special way of walking. You can see evidence for their different ways of walking if you compare their skeletons . . .

The skull

human being

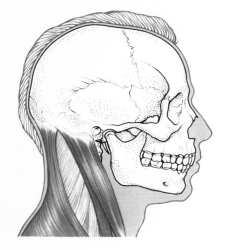

gorilla

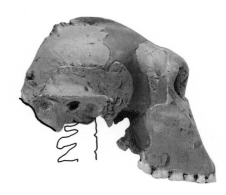

Compare these two australopithecine skulls with the human and gorilla skulls. Notice where the skull was joined to the backbone. And look at the marks where the neck muscles joined the skull. Do you agree that the australopithecine skulls are more like the human skull than the gorilla skull?

This is one piece of evidence that the australopithecines walked upright. We can find further evidence in other parts of the skeleton.

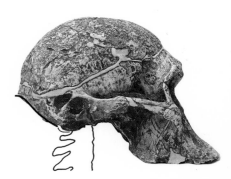

Australopithecine skull from Olduvai Gorge, Tanzania

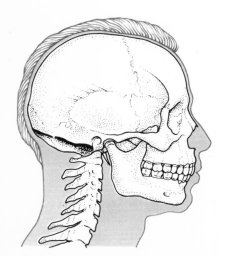

Human beings walk upright. Their heads are well balanced on top of their backbones. The backbone meets the skull towards the middle. The neck muscles that support the head are small and attached quite low down on the skull.

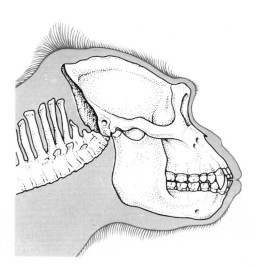

The backbone of a gorilla meets the skull towards the back, and the head needs much more support. The neck muscles are large and strong and are attached fairly high up on the skull.

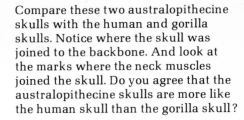

Australopithecine skull from Sterkfontein, South Africa

The hip

human being

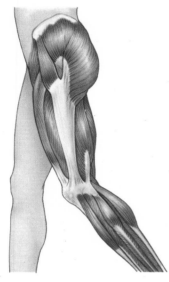

gorilla

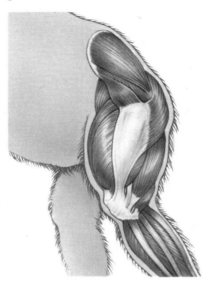

The hip is important in walking, because most of the muscles used in walking are attached to it. The relative size of some of these walking muscles

is quite different in human beings and gorillas. And the shape of the hip bone reflects these differences.

Look at this australopithecine hip bone. You can see that it is short and broad like a human hip bone, not long and narrow like a gorilla hip bone.

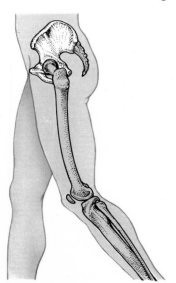

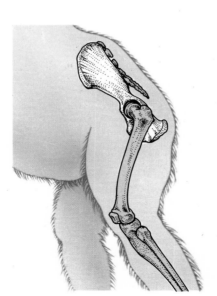

Australopithecine hip bone from Sterkfontein, South Africa.

The thigh bone

The thigh bone is also important in walking. Again, many of the muscles used in walking are attached to it. Compare this australopithecine thigh bone with the human and gorilla thigh bones. Even though the australopithecine thigh bone is incomplete, you can see that it is much more like the human thigh bone – it has a long neck, and the shaft is long and slightly curved.

The foot

Foot bones are small and fragile, and are not often preserved as fossils. But, a few australopithecine foot bones have been found. The bones from several different individuals have been used to reconstruct an australopithecine foot.

Compare this australopithecine foot with the feet of a human being and a gorilla. Notice that the toes are short like ours, and the big toe points forward. The gorilla's toes are very long, and its big toe sticks out at the side.

So the australopithecines seem to have had feet more like ours than like the gorilla's.

What kind of footprints do you think they made?

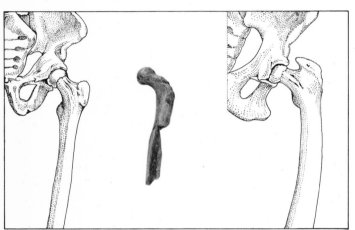

human being

Australopithecine thigh bone from Koobi Fora, Kenya.

gorilla

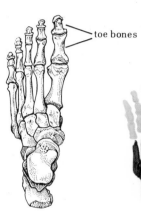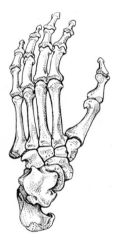

toe bones

human being

Australopithecine foot based on foot and big toe bones from Olduvai Gorge, Tanzania

gorilla

Lucy

In 1974, the remains of a skeleton about 3 million years old were found at Hadar in Ethiopia. Nearly half the skeleton had been preserved.

This skeleton, which is probably female, was nicknamed 'Lucy'. Lucy has since been named as a new kind of australopithecine, *Australopithecus afarensis* (Afar is the region of Ethiopia in which she was found).

Look carefully at this photograph of Lucy. See if you can decide whether or not she walked upright like a human being. Look especially at the hip bone and the thigh bone.

All the evidence from the different parts of the skeleton suggests that Lucy – and all the other australopithecines – must have walked upright, like human beings.

Fossilized footprints found at Laetoli, Tanzania. Made 3.5 million years ago by creatures that walked on two legs – australopithecines?

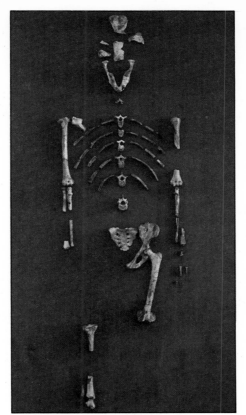

'Lucy', incomplete skeleton of an australopithecine from Hadar, Ethiopia.

But the australopithecines may not have walked in quite the same way as human beings. The blade of an australopithecine hip bone seems to have been more at the back of the body than at the side – in a human being it is at the side. And the neck of an australopithecine thigh bone is very flat, not rounded like the neck of a human thigh bone. These two differences suggest that the australopithecines probably did not stride along in the same way that we do.

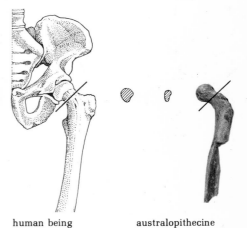

human being   australopithecine

Different kinds of australopithecines

The australopithecines were not all alike. We have found the remains of several different kinds.

Robust australopithecines

These two australopithecines belong to a group known as the **robust** australopithecines. They have several characteristics that human beings do not share.

Look at these teeth. The robust australopithecines have very large chewing teeth, but small canines and biting teeth. They were probably vegetarians, and their teeth are specially adapted to their vegetarian diet.

The specialized teeth of the robust australopithecines are unique characteristics that we do not share. So, although these australopithecines are our fossil relatives, they cannot be our direct ancestors.

Gracile australopithecines

Another group of australopithecines – the **gracile** or slender australopithecines – seem to have no unique characteristics of their own. This makes it difficult to decide how they are related to human beings.

Part of the skull of a robust australopithecine from Olduvai Gorge, Tanzania.

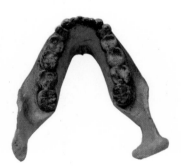

Jaw of a robust australopithecine from Peninj, Tanzania.

Part of the skull of a gracile australopithecine from Sterkfontein, South Africa.

Part of the skull of a robust australopithecine from Swartkrans, South Africa.

Jaw of a robust australopithecine from Swartkrans, South Africa.

Jaw of a gracile australopithecine from Sterkfontein, South Africa.

Are the gracile australopithecines more closely related to human beings? Or are they more closely related to the robust australopithecines?

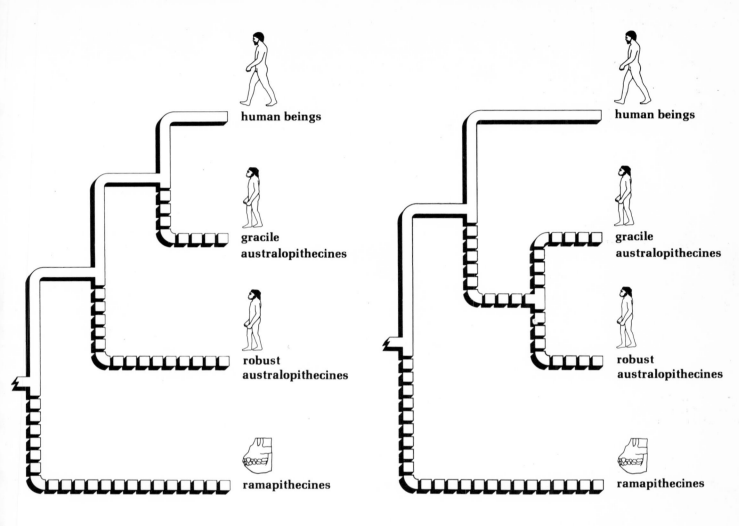

human beings

gracile australopithecines

robust australopithecines

ramapithecines

human beings

gracile australopithecines

robust australopithecines

ramapithecines

If this relationship is correct, we would expect to find some characteristics that gracile australopithecines and human beings share, but the robust australopithecines do not.

If this relationship is correct, we would expect to find some characteristics that the gracile and robust australopithecines share, but human beings do not.

At present, there is not enough evidence to decide.

Chapter 5
Man makes tools

The habilines

Another group of creatures thought
to be our fossil relatives lived in East
Africa from about 2 million to 1·5
million years ago. They have been
called the **habilines**, and they were
probably the first creatures to make
tools.

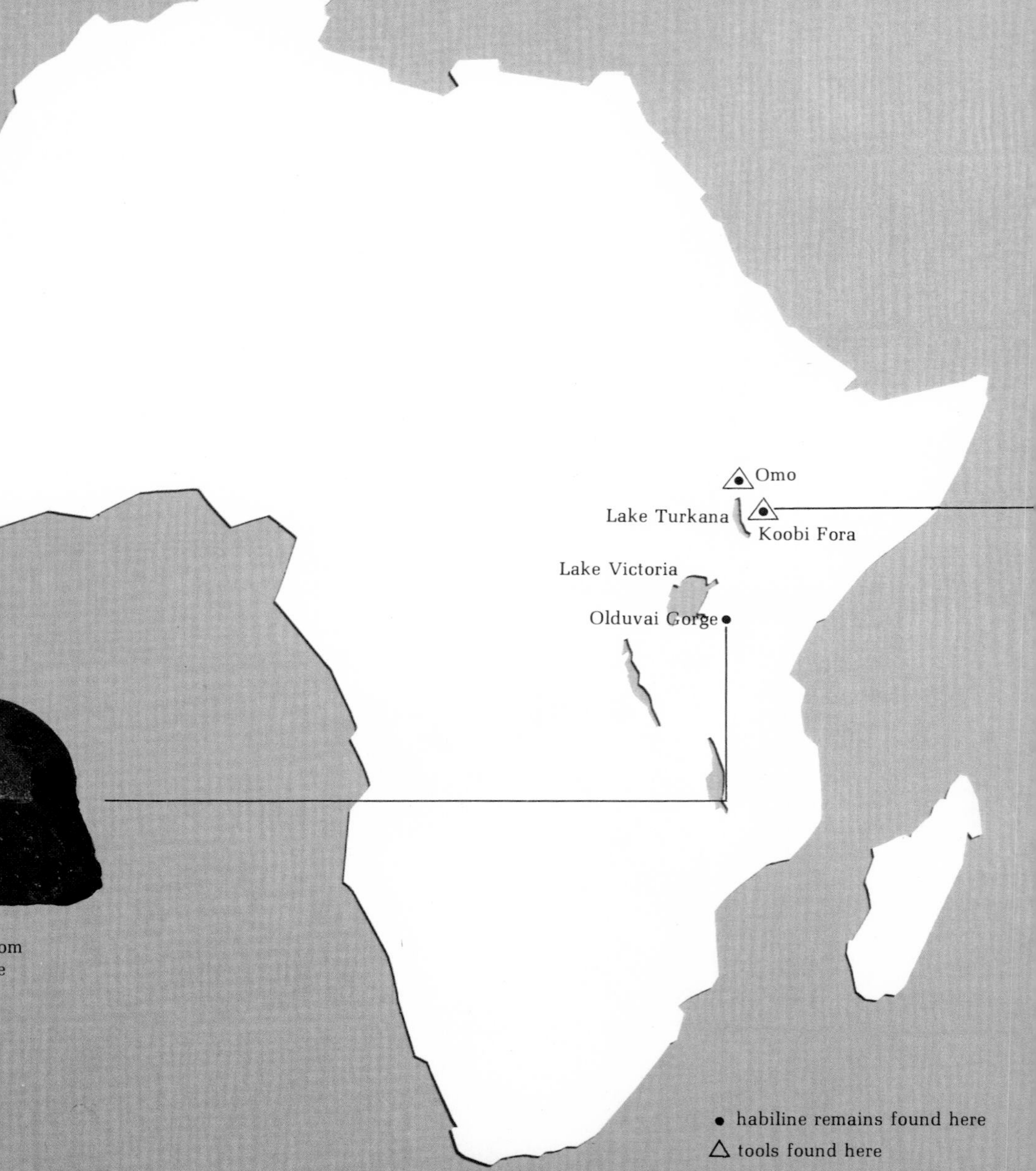

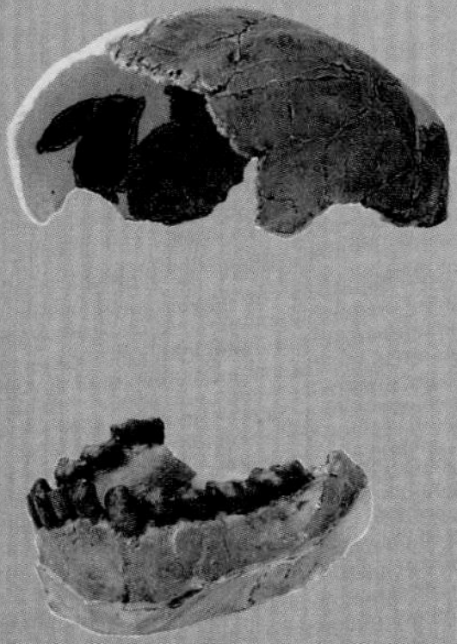

Omo

Lake Turkana

Koobi Fora

Lake Victoria

Olduvai Gorge

Fragments of the skull
of a habiline from
Olduvai Gorge,
Tanzania.

Pebble tool from
Olduvai Gorge

● habiline remains found here

△ tools found here

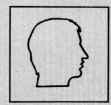

Part of the skull of a habiline from Koobi Fora, Kenya.

Flake tool and pebble tool from Koobi Fora.

How are the habilines related to man?

The habilines are closely related to human beings and the australopithecines. They share the same important characteristics:

- a short muzzle

- walking on two legs

You can find evidence for both these characteristics if you look closely at this skull.

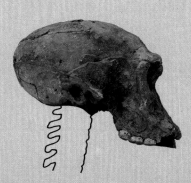

Part of the skull of a habiline from Olduvai Gorge, Tanzania.

Are the habilines more closely related to the australopithecines?

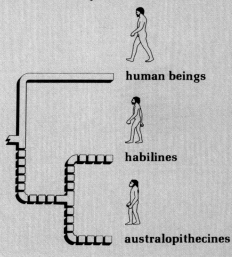

human beings

habilines

australopithecines

Or are they more closely related to human beings?

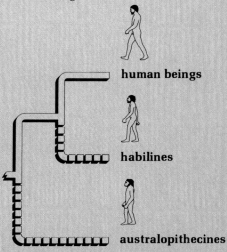

human beings

habilines

australopithecines

We think that the habilines are more closely related to human beings than to the australopithecines because there is evidence that they had large brains, and they made tools.

Man has a large brain

A large brain is another important human characteristic. Of course, there are other, much larger, animals that have even bigger brains. But no other animals of the same size have such a large brain.

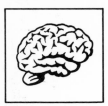

For example, a 60 kilogram chimpanzee has a much smaller brain than a 60 kilogram human being has.

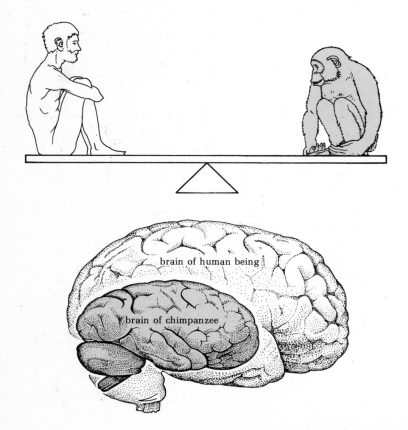

brain of human being

brain of chimpanzee

How do we know the brain size of a fossil?

Of course, we cannot measure the brains of fossils directly, because they are never preserved. But we can get some idea of how big the brain of a fossil was if we make a cast of the inside of its skull, or even part of its skull.

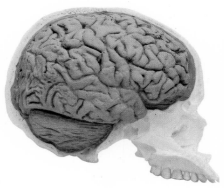

human skull with model of brain

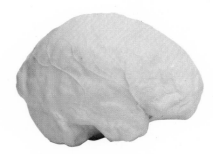

cast of the inside of a human skull

Such a 'brain cast' is very slightly larger than the brain itself, because there is a space between the skull and the brain.

The habilines had large brains

Although the habiline brain cast is much smaller than the brain of a human being, the habilines were much smaller creatures. From the size of their bones, we can estimate that they probably weighed only about 40 kilograms – about the same as a twelve-year-old child. So, for their size, the habilines had fairly large brains.

Of course, the australopithecines were also small. And, for their size, they also had quite large brains. But we think that the difference between the habiline and the australopithecine brain sizes is important because the habilines made tools. (You will find out more about this later in the chapter.)

Man makes tools

Human beings are the only living creatures that manufacture tools – we **use** tools to **make** tools.

Other animals, including chimpanzees and some birds, may select objects to **use** as tools. They may even modify the objects to make them into tools. But they have never been known to use tools to make tools as we do.

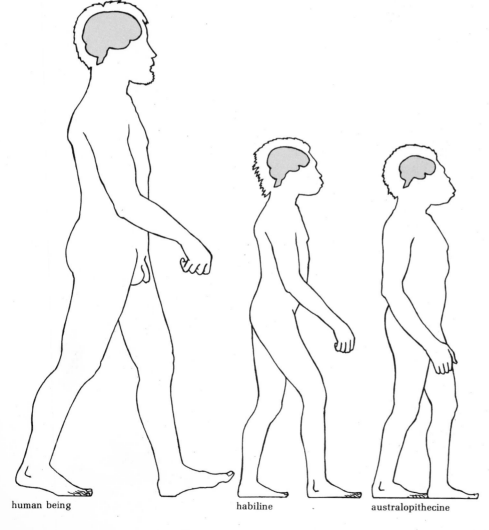

human being   habiline   australopithecine

Making tools is very different from merely using them. It involves solving problems . . .

No other living creature shares man's tool-making behaviour – it is a unique human characteristic.

The habilines made tools

The habilines were probably the first creatures to make tools. There is evidence that they collected stones, often from many kilometres away, and used other stones to reshape them into tools.

They made different tools for different purposes. Their 'toolkit' includes flakes and knife-like tools, as well as a variety of choppers and scrapers.

We can guess that the australopithecines may also have used stones and bones as tools, but there is no evidence that they made tools as the habilines did.

The remains of a habiline camp at Olduvai Gorge, Tanzania.

60

The habiline 'toolkit' found at
Olduvai Gorge is usually referred to as
the *Oldowan industry*. It includes:

Cutting tools
A stone flake held
between the thumb
and fingers makes a
good cutting tool.

Tools to make tools
A hammerstone like
this can be used for
striking flakes off
another stone to make
a tool with a sharp edge.

Scraping tools
We do not know what
these tools were used
for, but we can guess
that they may have
been used for scraping
meat from bones or for
scraping fat from skins.

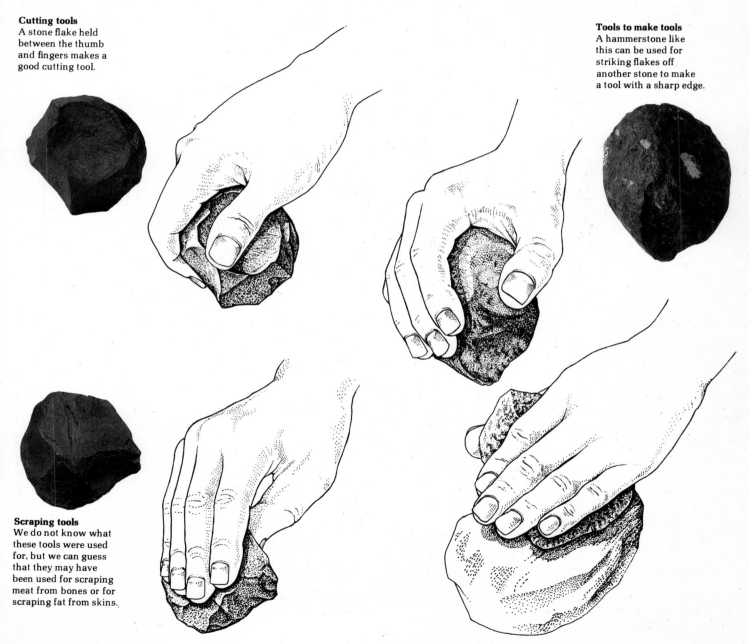

The habiline way of life
In search of food, some habilines find a small antelope.

They kill the antelope.

Someone makes tools from the
stones he was carrying with him.

They butcher the antelope
using the tools they have made.

They share out the meat. There is enough meat on the carcass to feed the whole group.

Then they move on, leaving the tools and bones behind them.

Were the habilines human?

Yes. Toolmaking is such an important human characteristic that we can think of the habilines as human beings.

Obviously toolmaking is very different from the other characteristics that we have used to test relationships. It is not a physical characteristic at all. It is a characteristic of **behaviour**, and it can be learned.

But it takes a great deal of intelligence to solve problems by making tools, and this is why we think that toolmaking is so important. The habilines' varied toolkit is good evidence that they had quite clever brains.

The scientific name for all living human beings is *Homo*, which is Latin for 'man'. We have given this name to the habilines too.

But the habilines are so different from us that we can think of them as a different kind of human being. So we have given them their own scientific name – *Homo habilis*, which means 'handy man'.

Modern human beings are called *Homo sapiens*, which means 'wise man'. In order to distinguish them from fossil human beings like the habilines, we shall refer to modern human beings as **modern man** in the rest of the book.

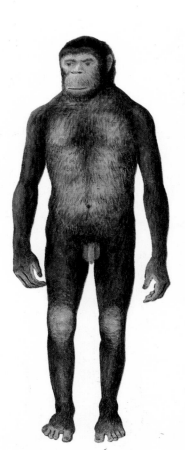

Habiline
Homo habilis

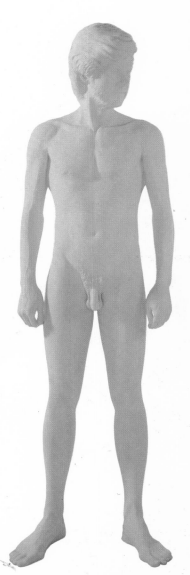

Modern man
Homo sapiens

Chapter 6

Man uses fire

The *Homo erectus* people

Remains of other fossil human beings have been found in Africa and South East Asia in rocks more than a million years old. These fossils have been given the name *Homo erectus* – 'upright man'.

Later *Homo erectus* people spread to Northern Asia and to Europe. There is evidence that these people used fire – which would certainly have helped to solve the problem of how to keep warm in these cooler northern climates.

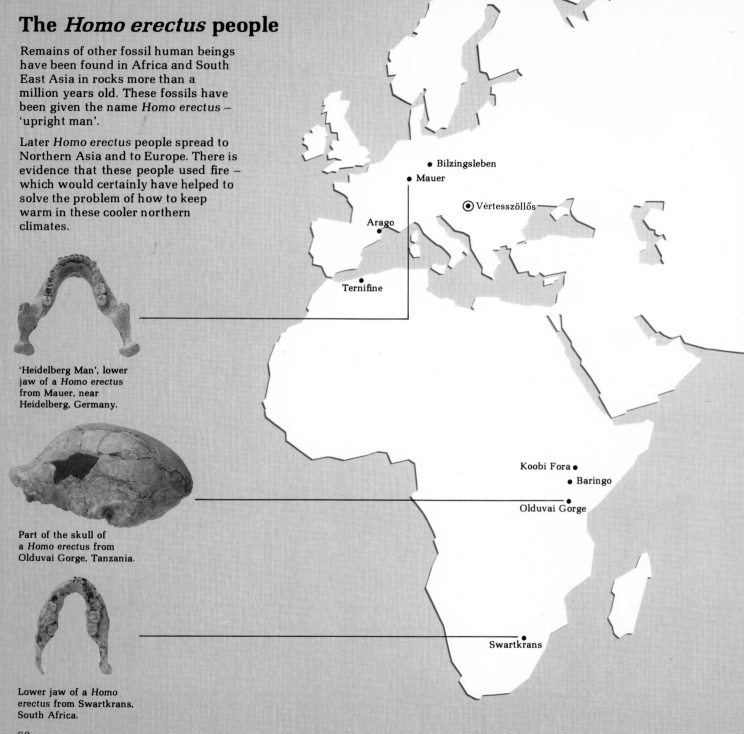

'Heidelberg Man', lower jaw of a *Homo erectus* from Mauer, near Heidelberg, Germany.

Part of the skull of a *Homo erectus* from Olduvai Gorge, Tanzania.

Lower jaw of a *Homo erectus* from Swartkrans, South Africa.

- Bilzingsleben
- Mauer
- Vértesszöllős
- Arago
- Ternifine
- Koobi Fora
- Baringo
- Olduvai Gorge
- Swartkrans

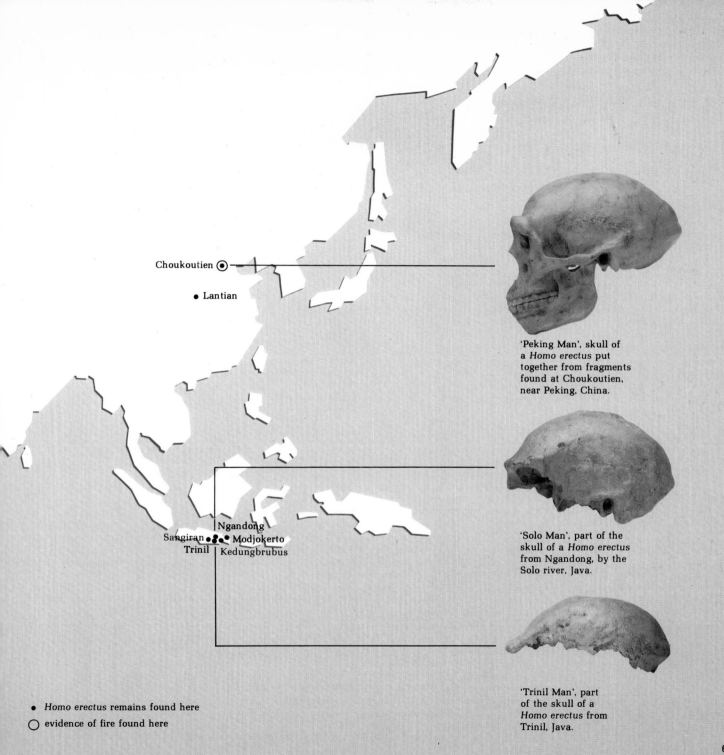

Choukoutien ◉

● Lantian

Ngandong
Sangiran ●●●● Modjokerto
Trinil | Kedungbrubus

'Peking Man', skull of
a *Homo erectus* put
together from fragments
found at Choukoutien,
near Peking, China.

'Solo Man', part of the
skull of a *Homo erectus*
from Ngandong, by the
Solo river, Java.

'Trinil Man', part
of the skull of a
Homo erectus from
Trinil, Java.

● *Homo erectus* remains found here
◯ evidence of fire found here

69

How are the *Homo erectus* people related to modern man?

If the *Homo erectus* people are closely related to modern man and the habilines, they must share the same characteristics . . .

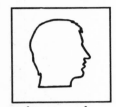

• a short muzzle

• walking on two legs

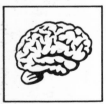

• a large brain

• tool-making

You can find evidence for most of these characteristics in these *Homo erectus* remains found at Trinil in Java . . .

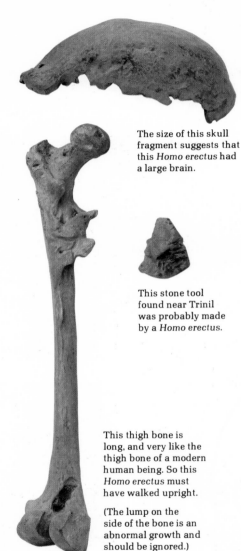

The size of this skull fragment suggests that this *Homo erectus* had a large brain.

This stone tool found near Trinil was probably made by a *Homo erectus*.

This thigh bone is long, and very like the thigh bone of a modern human being. So this *Homo erectus* must have walked upright.

(The lump on the side of the bone is an abnormal growth and should be ignored.)

So the *Homo erectus* people must be closely related to modern man and the habilines. But which are they **more** closely related to – modern man or the habilines?

Like modern man, the *Homo erectus* people used fire. And, for their size, they had larger brains than the habilines.

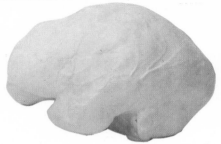

brain cast of a *Homo erectus*

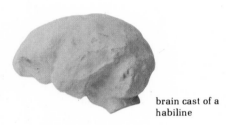

brain cast of a habiline

They also had smaller teeth – their wisdom teeth in particular were small and very like our own. All this suggests that the *Homo erectus* people are more closely related to modern man than to the habilines . . .

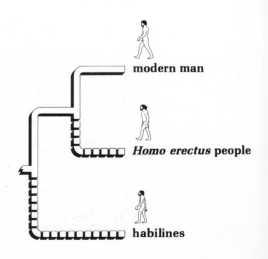

modern man

Homo erectus people

habilines

Man uses fire

Fire using is an important
characteristic of modern man.
We use fire to solve many different
kinds of problem.

We use fire for . . .

Using fire to solve problems is similar
to making tools to solve problems. It
requires intelligence and imagination,
and can be seen as evidence of a fairly
advanced brain. No other living
creature uses fire — it is a unique
characteristic of man.

heating

cooking

making things

The *Homo erectus* people used fire

There is evidence that the *Homo erectus* people who lived at Choukoutien near Peking in China were able to use fire. Ash layers, charcoal and burnt bones have been found with *Homo erectus* remains in a cave site at Choukoutien.

Ash layer
From burnt-out fires

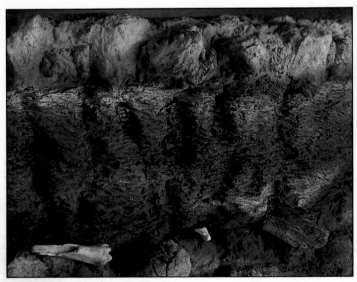

Evidence for fire at Choukoutien
Reconstruction based on information from the excavations at Choukoutien,

Charcoal
The remains of burnt-out fires

Scorched stone
Hot stones may have been plunged into water to heat it for cooking purposes

Burnt bone
From a deer that had probably been killed and eaten

The excavation site at Choukoutien near Peking, China.

Shells of hackberry nuts
These nuts were probably crushed and eaten

The *Homo erectus* way of life

We have evidence that the *Homo erectus* people used fire, and we can imagine how this may have influenced their way of life...

A fire would provide a centre for a permanent home base. People would gather round the fire, and so would tend to live in groups.

Some members of the group would band together to go and hunt large animals. Others would gather seeds and fruits. Food would be brought back to the home base to be cooked and shared out amongst the members of the group.

Archaeologists are looking for evidence of such a way of life at the various sites where the remains of *Homo erectus* people, or their tools, have been found.

Not our direct ancestors

The *Homo erectus* people were not quite like us. Compare this reconstructed *Homo erectus* skull with the modern human skull. You will see that the *Homo erectus* skull has several characteristics that the modern skull does not share. Because of these special characteristics, we think that the *Homo erectus* people were not our direct ancestors.

Homo erectus skull, 'Peking Man'

modern human skull

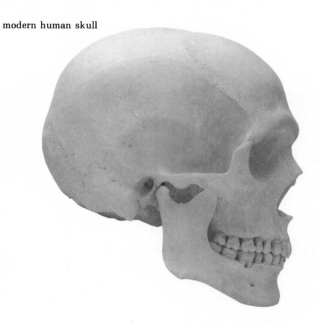

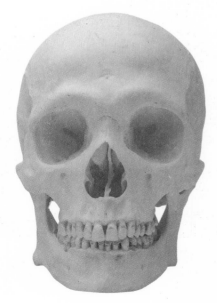

thick ridge along back of skull

keeled top of skull —

straight, thick brow ridge —

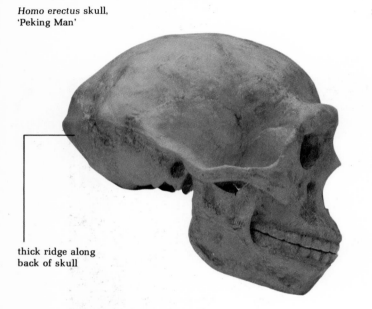

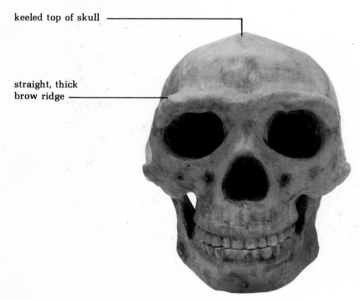

What shall we call this fossil?

Some of the more recent *Homo erectus* fossils, like this skull from Petralona in Greece, show evidence of such a large brain that some people think they should be grouped with modern man in the species *Homo sapiens*. But, in this book, we have called them *Homo erectus* because they share the unique *Homo erectus* characteristics illustrated on the opposite page.

Names like *Homo erectus* are useful, because they help us to talk about **groups** of fossils, rather than about individual specimens. But they are only labels, and can cause problems when a fossil – like the Petralona skull – shows characteristics of more than one group.

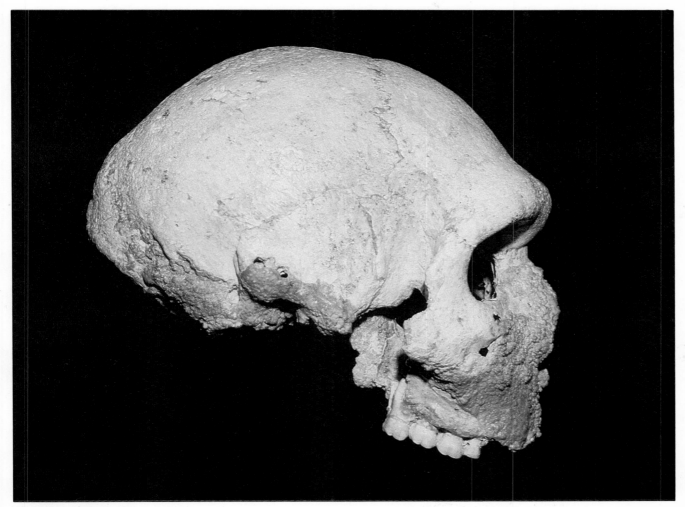

Homo erectus skull from Petralona, Greece.

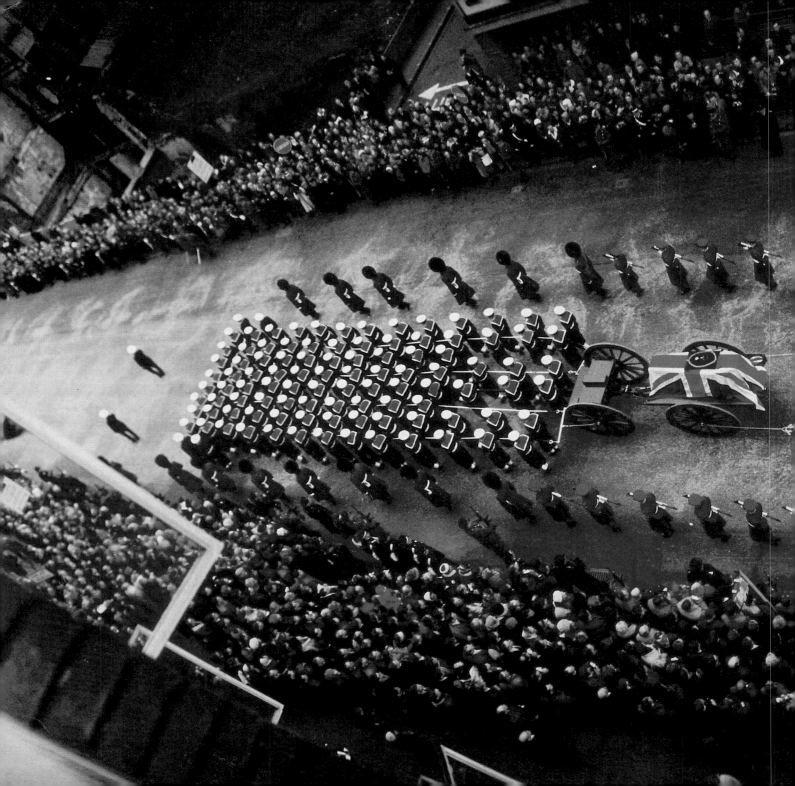

Chapter 7
Man has ceremonies

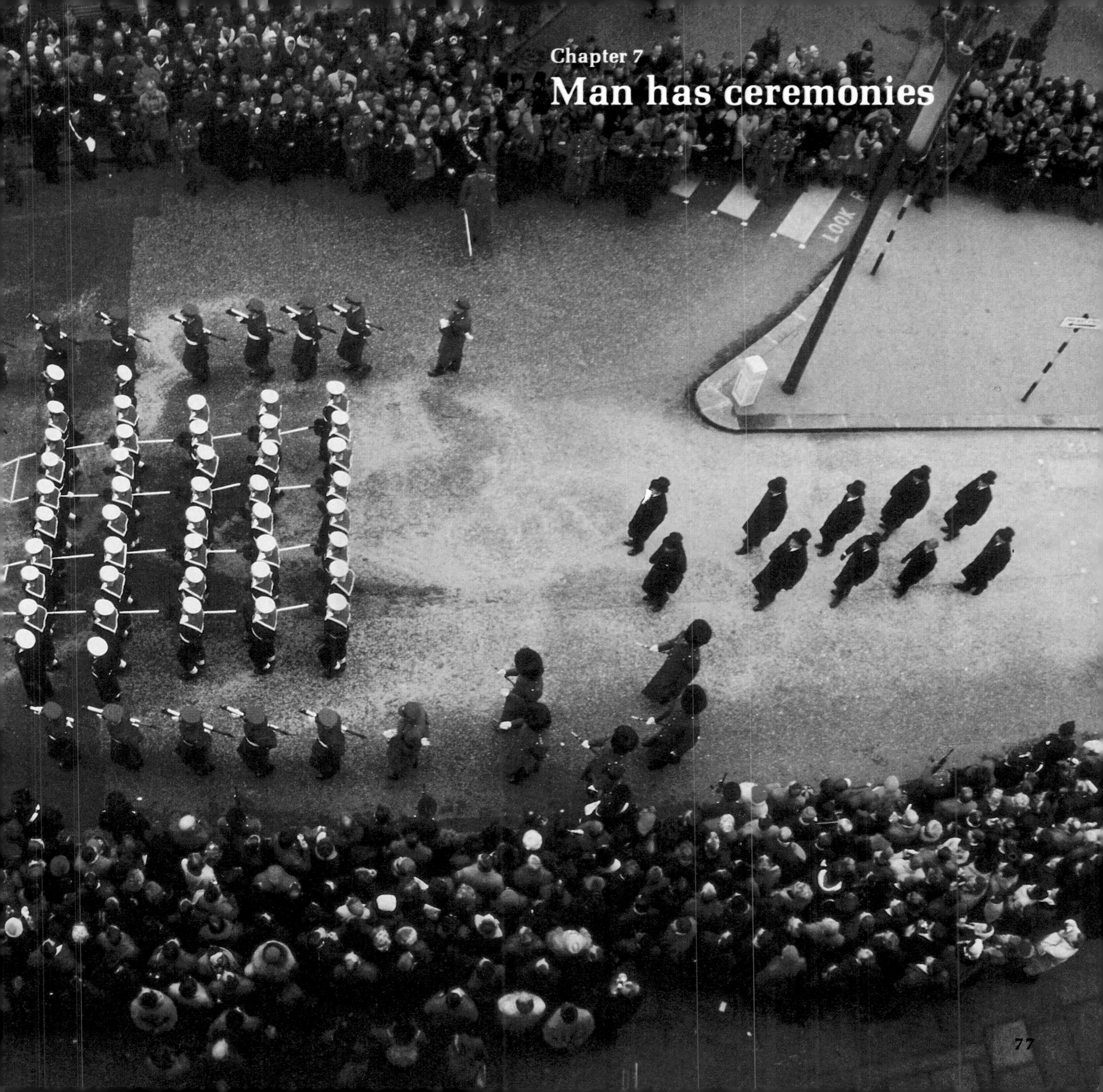

The neandertals

The neandertals lived in Europe and the Middle East before and during the last Ice Age. Most of them lived between 100 000 and 40 000 years ago. They were probably the first people to have ceremonies for their dead.

They were given the name **neandertal** (sometimes spelt neanderthal) because the first specimen to be scientifically described was found in a cave in the Neander Valley – *Neander Thal* in Old German.

Part of the skull of a neandertal from the Neander Valley, Germany. The first specimen to be given the name neandertal.

Part of the skull of a neandertal found at Forbes Quarry, Gibraltar in 1848. The first neandertal to be found.

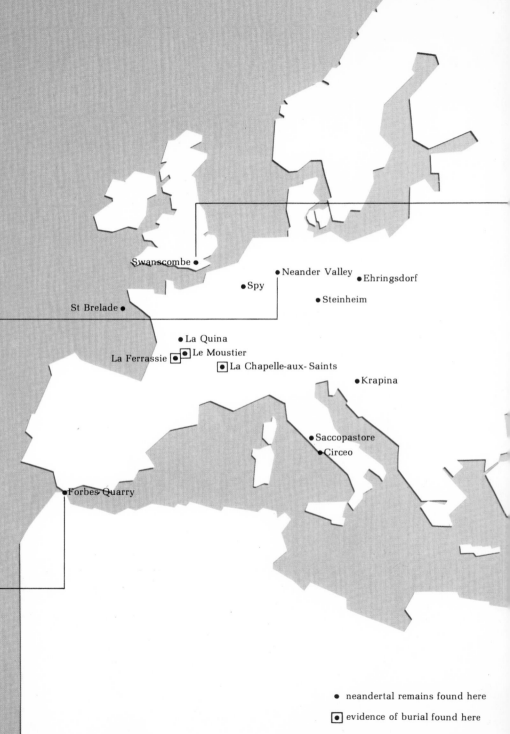

Swanscombe ●

● Neander Valley

St Brelade ●

● Spy

● Ehringsdorf

● Steinheim

● La Quina

La Ferrassie [⊙] [⊙] Le Moustier

[⊙] La Chapelle-aux-Saints

● Krapina

● Saccopastore

● Circeo

● Forbes Quarry

● neandertal remains found here

[⊙] evidence of burial found here

Teshik-Tash ◉

◉ Kiik-Koba

◉ Shanidar

Amud ◉

Tabūn ◉

The earliest known Briton.
'Swanscombe Man', three
skull fragments from
Swanscombe, England.

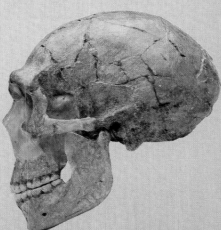

Skull of a neandertal
from Amud Cave, Israel.

79

How are the neandertals related to modern man?

During excavations at Tabūn in Israel, complete neandertal skeletons were found. Flake tools and burnt bones were also found there – providing evidence that the neandertals made tools and used fire.

Flake tool found with neandertal skeleton at Tabūn

Burnt bone found with neandertal skeleton at Tabūn

This neandertal woman is based on a skeleton found at Tabūn. (At the end of the chapter, you can find out how she was made.) You can see that she has a short muzzle and walked upright. And the size of her head suggests that she had a large brain.

So the neandertals must be closely related to modern man and the *Homo erectus* people, because they share the same characteristics. But which are they **more** closely related to – modern man or the *Homo erectus* people?

The neandertals have large brow ridges, like the *Homo erectus* people have.

Homo erectus skull, 'Peking Man'

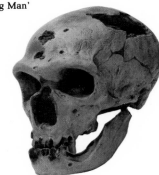

Neandertal skull from La Chapelle-aux-Saints, France

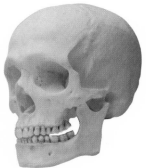

Modern human skull

So they could be closely related to the *Homo erectus* people . . .

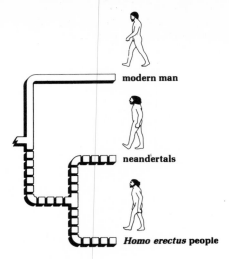

modern man

neandertals

Homo erectus people

But like modern man, the neandertals had very large brains – most of them had a brain larger than 1300 ml. And, like modern man, they buried their dead. So we think that the neandertals are more closely related to modern man . . .

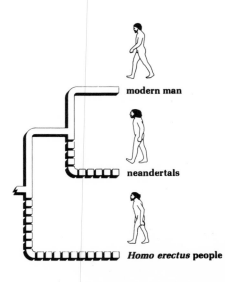

modern man

neandertals

Homo erectus people

80

Homo sapiens

Compare this neandertal skull with a modern human skull. The most obvious difference is in the general shape of the face. But there are other differences . . .

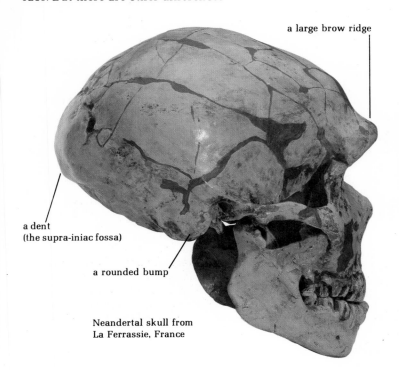

a large brow ridge

a dent
(the supra-iniac fossa)

a rounded bump

Neandertal skull from
La Ferrassie, France

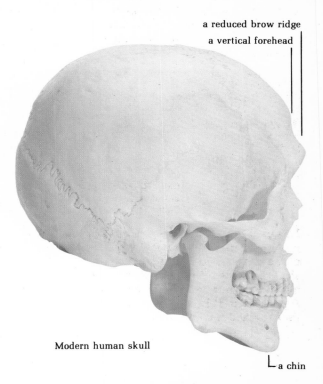

a reduced brow ridge

a vertical forehead

Modern human skull

a chin

There are also differences in other parts of the skeleton. **But** because the neandertals and modern man share these two important characteristics . . .

- an average brain size of 1330ml
- burial of the dead – ceremonies

. . . the differences are considered unimportant, and the neandertals and modern man are grouped together in the same species – *Homo sapiens*.

The neandertals' unique characteristics are recognized by calling them a subspecies of modern man – *Homo sapiens neanderthalensis*. They were an ancient form of *Homo sapiens* that died out about 40 000 years ago.

Man has ceremonies

Ceremonies are an important characteristic of modern man. We live in societies, and we recognize ourselves as part of a society when we take part in its ceremonies.

We have ceremonies ...

Ceremonies like these reflect our complex social structure, and are an important part of human culture. Like tool-making and the use of fire, ceremonies provide indirect evidence of man's intellectual – and cultural – development.

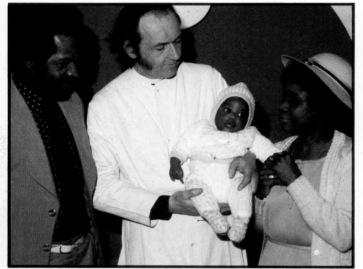

when a baby is born

when two people are married

when someone dies

82

The neandertals had ceremonies

The neandertals were probably the first people who had ceremonies for their dead. There is evidence for such ceremonies at several sites where neandertal remains have been found.

These are the remains of a neandertal child found buried at Teshik Tash in the USSR. The body was placed in a shallow pit, and pairs of goat horns were placed in a circle around it. A fire was lit beside the grave.

Reconstruction of
a neandertal burial,
based on remains found
at Teshik-Tash, USSR

A neandertal from Kent?

These are the remains of the earliest known Briton. They are perhaps 250 000 years old and were found at Swanscombe in Kent, not far from London.

The three skull bones were discovered at separate times (in 1935, 1936 and 1955). But they clearly come from the same person – a young woman.

On the back of her skull, there is the small dent (the supra-iniac fossa) that is one of the special characteristics of the neandertals. So the Swanscombe people may have been some of the earliest neandertals.

Other evidence of these ancient inhabitants of Britain has also been found in the river gravels at Swanscombe . . .

Part of a deer antler
Probably from an animal that had been killed and eaten.

Fragment of elephant bone
Probably from an animal that had been killed and eaten.

Flint hand axe
Hand axes like this have a long cutting edge, and make better cutting tools than stone flakes do.

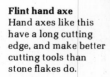

84

From these remains, we can begin to work out how the Swanscombe people lived ...

The amount of 'rubbish' (broken bones and used stone tools) suggests that the Swanscombe people may have camped here for several months – perhaps for a whole winter.

They hunted mainly deer – a kind of fallow deer.

There were also elephants and rhinoceroses living in the area at the time, and the Swanscombe people may occasionally have found these animals trapped in mud or marshy ground. If they did, they probably butchered them and shared out the meat.

Reconstructing a neandertal

For the exhibition of **Man's place in evolution**, one of the Natural History Museum's modelmakers produced a unique reconstruction of a neandertal woman. He based her on the 41 000 year old remains of a female neandertal skeleton, found at Tabūn, near Mount Carmel in Israel.

The neandertal woman took about three months to reconstruct. First of all, the modelmaker used measurements of the Tabūn bones to work out her height, the relative lengths of her limbs, and so on. Then, using his own

knowledge of anatomy, together with published information about neandertals, and advice from Museum scientists, he produced **drawings** to show what the reconstruction would look like.

When the drawings had been approved, and the model's exact pose had been decided on, the modelmaker could begin to plan the **armature** – the supporting framework on which the model would be built. He made the armature from wood and steel, and added a plaster cast of the original skull from Tabūn.

Polystyrene was used to form the bulk of the woman's torso, and the rest of the body was built up from layers of scrim (sacking) soaked in plaster. Later, a layer of plasticene was added. The modelmaker used the plasticene to create all the fine details – the exact contours of the woman's flesh, her skin texture, hair and so on.

1 Research and drawing

2 Welding the armature

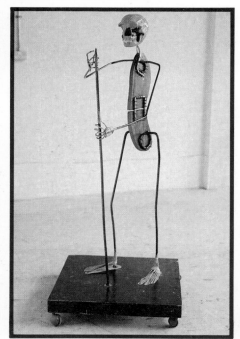

3 The completed armature

4 Building up the model with scrim and plaster

5 Modelling the details in plasticene

6 The completed 'plasticene' model

When the model had been completed to everyone's satisfaction, the modelmaker made a **mould** from it. Because of the model's complex shape, the mould had to be carefully planned, and was made in seven separate sections. Each of these was built up on the model from layers of liquid silicon rubber, strengthened with pieces of cloth. Each section was then encased in a strong 'jacket' of glass-reinforced plastic (GRP). When all the sections were complete, they were bolted together around the model to ensure a perfect fit.

The mould was then taken apart and the inside of each section was painted with a 'gel coat' of polyester resin. Specially-made glass eyes were put in place, and the mould was lined with layers of GRP. While this was still wet, the pieces were bolted back together again.

When the GRP was dry, the mould was taken apart. Inside, the GRP had formed a strong, hollow, lightweight replica or **cast** of the original model.

This cast was carefully cleaned and all the 'seam lines' were removed. Finally, the neandertal woman was painted, ready for display.

7 Making the mould

8 Applying the 'gel coat' to the mould

9 The finishing touches

Modern man

By about 30 000 years ago modern man had spread to nearly all parts of the world. Physically, they were just like people living today.

Man the hunter

These early people lived mainly by hunting large animals such as deer, horses and mammoths. In order to kill these animals, they must have lived and hunted in large organized groups.

Each group would probably have had a leader. And the work of hunting, making traps, building shelters and caring for children would have been divided amongst members of the group.

This kind of social organization could best be achieved if the members of the group could communicate with each other using **language**.

Can we find any evidence that they did use language?

Modern man uses symbols

Language is an important characteristic of modern man. In language, we use sounds – words – to represent things and ideas. In a similar way we use many other kinds of symbols to represent things, ideas and feelings.

We use symbols in

language

music

art

Man the hunter used symbols

The hunters of the last Ice Age used symbols too. They represented the animals they hunted in magnificent cave paintings. They carved female figures out of stone and bone, or moulded them in clay. They even made simple musical instruments.

There is no way of knowing when man first began to use language. We cannot tell from the fossils. But we can be fairly sure that the people who made these paintings and carvings also used language.

Art and language are closely linked. They both involve the use of symbols to represent things and ideas. This is why we think that the Ice Age hunters must have used language – their art is our evidence.

movement—gesture

Woman carved in soapstone, from Balzi Rossi in Italy. About 20 000 years old. (Actual size.)

mathematics

No other creature uses this wide range of symbols. The use of symbols in art, music, language and so on is a unique characteristic of modern man.

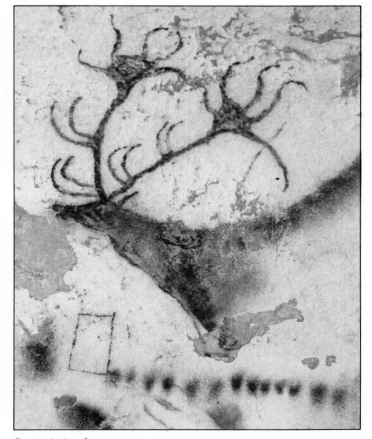

Cave painting, from Lascaux in France. About 17 000 years old. (Original about 1.4 metres wide)

Woman's head carved in ivory, from Brassempouy in France. About 18 000 years old. (Actual size.)

91

The development of farming

Towards the end of the last Ice Age, people began to farm in several different parts of the world. And, by about 7000 years ago, farmers already grew crops such as wheat and barley, and kept animals such as sheep, cattle and pigs.

Farming probably developed in response to a shortage of food. Towards the end of the Ice Age, the human population was steadily increasing. But climatic changes had led to the extinction or migration of many of the larger animals. So, in many areas, there were far fewer animals to hunt. Also, in dry areas, people would be forced to live near permanent supplies of water.

We can imagine that the people in these settled communities must have been forced to store food for hard times. They probably stored nourishing grains, and kept animals tethered as a kind of 'walking food store'. Such practices could gradually have developed into farming.

Maize was first cultivated in Central America.

Tehualán Valley ●

● Guila Naquitz Cave

Evidence of farming more than 7000 years ago:
- ● cultivated plants
- ○ domesticated animals

⌒ Fertile Crescent

Llamas were first domesticated in South America.

Uchcumachay ●

◉ Ayocucho Basin

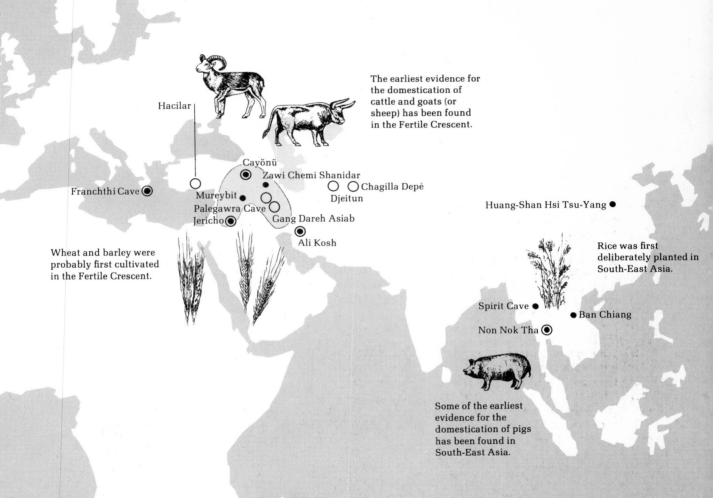

The earliest evidence for
the domestication of
cattle and goats (or
sheep) has been found
in the Fertile Crescent.

Hacilar

Franchthi Cave ◉

Cayönü

Zawi Chemi Shanidar ◉

Chagilla Depé ○

Djeitun

Mureybit

Palegawra Cave

Jericho ◉

Gang Dareh Asiab

Ali Kosh ◉

Huang-Shan Hsi Tsu-Yang ●

Wheat and barley were
probably first cultivated
in the Fertile Crescent.

Rice was first
deliberately planted in
South-East Asia.

Spirit Cave ●

Ban Chiang

Non Nok Tha ◉

Some of the earliest
evidence for the
domestication of pigs
has been found in
South-East Asia.

93

Man farms

Farming is another important
characteristic of modern man.
Farming involves . . .

storing crops

planting crops

harvesting crops

herding animals

feeding animals

breeding animals

A farmer must be able to plan for the future. He must store grain for planting next year, and he must store food to feed himself and his animals during the winter.

Farming does not require any new mental or physical skills. The Ice Age hunters would have needed the same amount of foresight, imagination and planning when they were hunting big game.

But farming *was* a very important step forward for modern man, because it was a solution to the problem of getting enough to eat. Farming produces more food from the same amount of land. So the communities that adopted farming would have had an advantage over the hunters. Their populations would have been able to increase, and these early farming settlements were the beginning of the towns and cities of the modern world.

so that people can eat.

Evidence of early farming

There is evidence that the people who lived at Jericho, near the Dead Sea, 9000 years ago were farmers.

Excavations show that there was a large settlement at Jericho – as many as 2000 people may have lived there. Such a large population could not have been supported by hunting alone.

Part of a house
excavated at Jericho

There is evidence for cultivated barley at Jericho. Grains of 2-row barley found there are larger than grains of wild 2-row barley, and are believed to come from cultivated plants.

'cultivated' two-row barley

wild two-row barley

This flint sickle from Jericho was evidently used for cutting ears of corn — its edge has been polished by the silica in their stalks. Sickles like this would be needed only where there was a lot of corn to cut.

reconstructed flint sickle

There is also evidence for domesticated goats. Early domesticated animals were usually tethered and sometimes poorly fed. Small-sized animals were better able to survive this treatment. So, at first, domesticated animals were smaller than wild animals of the same species.

bone of 'domesticated' goat from Jericho

bone of wild goat from Jericho

This quern (grindstone) from Jericho must have been used for grinding corn. Grindstones very like this are still in use in some parts of the world.

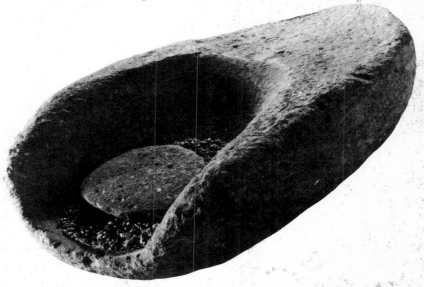

Chapter 9
Man has evolved

When we compare ourselves with our fossil relatives, we can find evidence that man has evolved.

When we look at the fossil remains, we find evidence of physical differences. For example we are taller than the first human beings – the habilines. And we have much larger brains.

There is also evidence that human behaviour has become much more complicated. For example the *Homo erectus* people used fire and, later, the neandertals took part in ceremonies for their dead.

The first modern people lived in organized groups and hunted large animals, which they depicted in their paintings. Some of their descendants lived in settled communities that developed first into farming villages, and then into the towns and cities of the modern world.

From the time of the first toolmakers, man has survived by controlling the environment in a way that no other animal has.

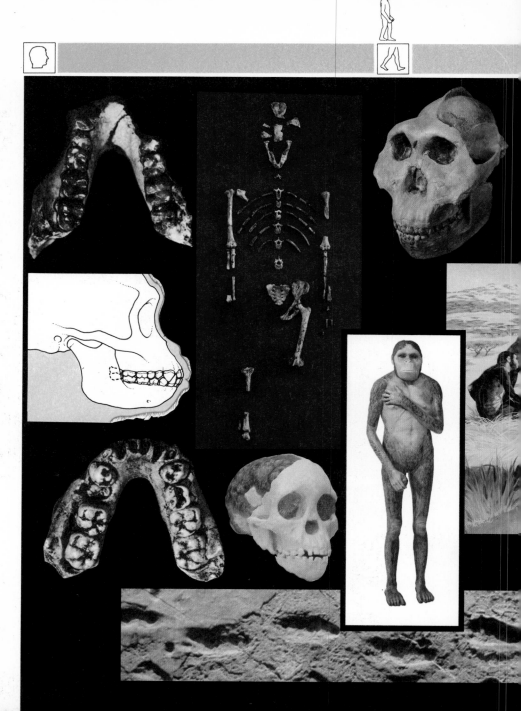

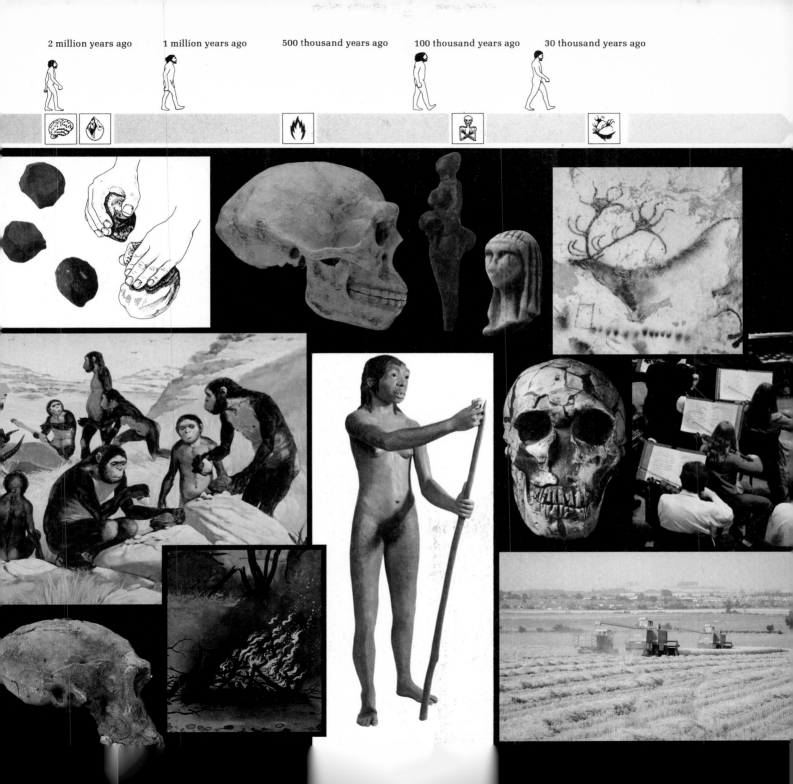

Man's place in evolution

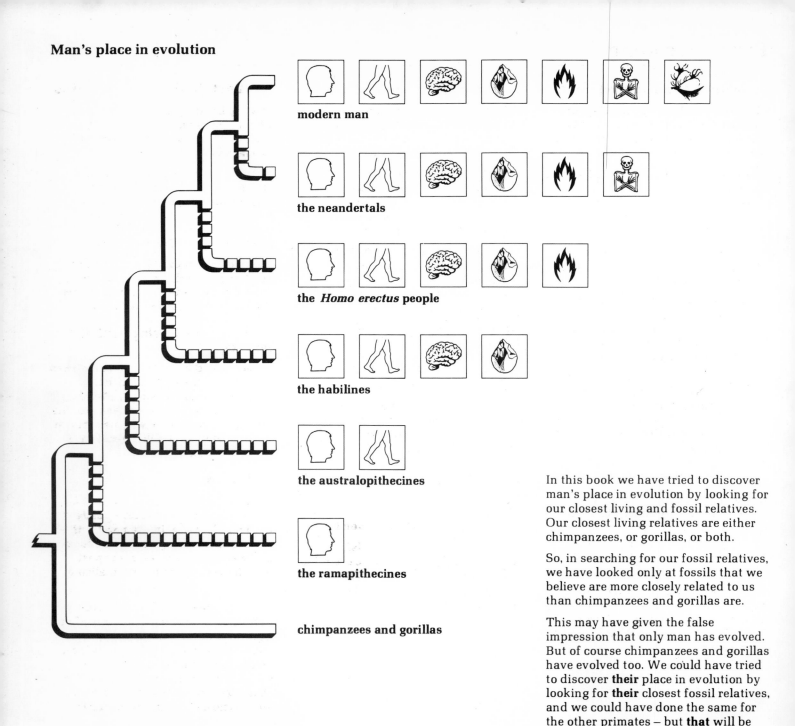

modern man

the neandertals

the *Homo erectus* people

the habilines

the australopithecines

the ramapithecines

chimpanzees and gorillas

In this book we have tried to discover man's place in evolution by looking for our closest living and fossil relatives. Our closest living relatives are either chimpanzees, or gorillas, or both.

So, in searching for our fossil relatives, we have looked only at fossils that we believe are more closely related to us than chimpanzees and gorillas are.

This may have given the false impression that only man has evolved. But of course chimpanzees and gorillas have evolved too. We could have tried to discover **their** place in evolution by looking for **their** closest fossil relatives, and we could have done the same for the other primates – but **that** will be another book . . .

Further reading

... about man's living relatives

Monkeys and Apes by Prue Napier, Hamlyn All-Colour Paperbacks, 1971. A colourful little book that contains an amazing amount of information.

The Primates by Sarel Eimerl and Irven DeVore, *Life Nature Library*, Time–Life International, 1966. A well-illustrated account of the biology and behaviour of primates.

The Chimpanzees by Prue Napier, *Bodley Head New Biologies*, Bodley Head, 1974. For children.

In the Shadow of Man by Jane van Lawick-Goodall, Collins, 1971. A fascinating first-hand account of life amongst wild chimpanzees on the shore of Lake Tanganyika.

Gorillas by Colin P. Groves, Arthur Barker/Arco Publishing Company, 1970. An affectionate study of the life and behaviour of gorillas, both in captivity and in their natural surroundings.

... about evolution

Evolution by Colin Patterson, British Museum (Natural History), 1978. An authoritative introduction to modern evolutionary theory for people who have little or no technical knowledge of biology.

Life on Earth by David Attenborough, Collins/BBC, 1979. A well-written account of the evolution of life on Earth, illustrated by excellent photographs.

... about man's fossil relatives

The Evolution of Early Man by Bernard Wood, Peter Lowe, 1976. A good background book: easy to read, with useful diagrams and pictures.

Early Man by F. Clark Howell, *Life Young Readers Library*, Time–Life International, revised edition, 1977. A well-illustrated and easy-to-read account of human evolution.

Man the Toolmaker by Kenneth P. Oakley, British Museum (Natural History), 6th edition, 1972. A good, concise introduction to the tools made by early man.

Origins by Richard E. Leakey and Roger Lewin, Macdonald and Janes, 1978. A well-illustrated book that gives an interesting and plausible account of early man.

The Emergence of Man by John E. Pfeiffer, Thomas Nelson & Sons, 3rd edition, 1978. A very readable textbook on human evolution.

An Introduction to the Study of Man by J. Z. Young, Oxford University Press, 1971. A comprehensive account of the biology, behaviour and evolution of man, for 6th formers and beyond.

The Descent of Woman by Elaine Morgan, Corgi, 1974. Aquatic evolution – an alternative view of 'man's' beginnings. Easy to read, and often amusing.

... about the discovery of fossils

Missing Links by John Reader, Collins, 1981. A very readable account of various 'fossil man' discoveries and the people who made them.

Pekin Man by H. Shapiro, Allen & Unwin, 1974. A description of the finding and subsequent loss of the Peking fossils, and an attempt to interpret the way of life of the *Homo erectus* people at Peking.

Meeting Prehistoric Man by G. H. R. von Koenigswald, Thames & Hudson, 1966. A very readable account of how the fossils from Java were found.

... about the development of farming

Domesticated animals from early times by Juliet Clutton-Brock, British Museum (Natural History), 1981. This new book looks at the evidence for the first domestication of mammals, and at the exploitation of wild animals now and in the past.

... for fun

Once Upon an Ice Age by Roy Lewis, Terra Nova Editions, 1979. A really good novel, combining fact and fiction. The reader can imagine the emotions experienced by 'hordes' of the past who discover fire and the tantalizing smells of cooked food.

The Inheritors by William Golding, Faber, 1973. A delightful fantasy about a group of neandertals meeting with another group of early human beings, and their reaction to their strange appearance, customs and rituals. The characters are enchanting – an easy way to learn about the neandertals.

Names of the fossils featured in this book

(The numbers in brackets are the catalogue or Museum numbers of these specimens.)

Ramapithecines

p.37, 44

From Çandır, Turkey
Ramapithecus wickeri, originally called *Sivapithecus alpani*.

p.37

From Chinji, Pakistan (GSI.D.118/119)
Ramapithecus punjabicus, originally called *Dryopithecus punjabicus*.

p.37

From Haritalyangar, India (YPM 13799)
Ramapithecus punjabicus, originally called *Ramapithecus brevirostris*.

p.37

From Hasnot, Pakistan (YPM 13814)
Ramapithecus punjabicus, originally called *Bramapithecus thorpii*.

p.37

From Fort Ternan, Kenya (FT 46, 47)
Ramapithecus wickeri, originally called *Kenyapithecus wickeri*.

p.39

From Potwar, Pakistan (GSP 4622/4857)
Ramapithecus punjabicus.

Australopithecines

p.42

From Taung, South Africa
Australopithecus africanus.

p.43

From East Turkana, Kenya (KNM-ER 406)
A robust australopithecine, referred to as *Australopithecus robustus boisei*.

p.43, 48, 52

From Olduvai Gorge, Tanzania (OH5)
A robust australopithecine, originally named *Zinjanthropus boisei*. Also referred to as *Australopithecus (Zinjanthropus) robustus boisei*.

p.43, 52

From Swartkrans, South Africa (SK48)
A robust australopithecine, originally named *Paranthropus crassidens*. Also referred to as *Australopithecus robustus*.

p.45

From Makapansgat, South Africa (MLD2)
Originally called *Australopithecus prometheus*. Variously referred to as *Australopithecus africanus*, *Homo africanus* and *Australopithecus robustus*.

p.48, 52

From Sterkfontein, South Africa (Sts 5)
A gracile australopithecine, *Australopithecus africanus*, originally called *Plesianthropus transvaalensis*.

p.49

Hip bone from Sterkfontein, South Africa (Sts 14)
A gracile australopithecine, *Australopithecus africanus*, originally called *Plesianthropus transvaalensis*.

p.50/51

Thigh bone from Koobi Fora, Kenya (KNM-ER 738)
Australopithecus species.

p.50

Footbones and big toe bone from Olduvai Gorge (OH8, OH10)
Originally called *Homo habilis*. Now often thought of as an australopithecine.

p.51

'Lucy' from Hadar, Ethiopia (AL 288-1)
Australopithecus afarensis

p.52

From Peninj, Tanzania
A robust australopithecine,
Australopithecus (Zinjanthropus) robustus boisei

p.52

From Swartkrans, South Africa (SK 23)
A robust australopithecine,
Australopithecus robustus. Also called
Paranthropus robustus crassidens.

p.52

Jaw from Sterkfontein, South Africa (Sts 52b)
A gracile australopithecine,
Australopithecus africanus, originally called *Australopithecus prometheus*.
Also called *Plesianthropus transvaalensis*.

Habilines

p.56

From Olduvai Gorge, Tanzania (OH7)
Homo habilis.

p.57

From Koobi Fora, Kenya
(KNM-ER 1813)
Referred to as *Homo habilis*.

p.57

From Olduvai Gorge, Tanzania (OH24)
Referred to as *Homo habilis*.

Homo erectus people

p.68

'Heidelberg Man' from Mauer, Germany
Homo erectus heidelbergensis,
originally called *Homo heidelbergensis*.

p.68

From Olduvai Gorge, Tanzania (OH9)
Homo erectus, originally called *Homo leakeyi*. Sometimes known as 'Chellean Man'.

p.68

From Swartkrans, South Africa (SK15)
Homo erectus, originally called *Telanthropus capensis*. Has also been called *Homo habilis*.

'Peking Man' from Choukoutien, China
Homo erectus pekinensis, originally
called *Sinanthropus pekinensis*.

p.69

'Solo Man' from Ngandong, Java
Homo erectus soloensis, originally
called *Javanthropus soloensis*. Has also
been named *Homo sapiens*.

p.69

'Trinil Man' from Trinil, Java
Homo erectus erectus, originally called
Pithecanthropus erectus.

p.75

From Petralona, Greece
Homo erectus heidelbergensis,
originally thought to be a neandertal.
Has also been called *Homo sapiens
heidelbergensis*.

Neandertals

p.78

From the Neander Valley, Germany
Homo sapiens neanderthalensis,
originally called *Homo neanderthalensis*.

p.78

From Forbes Quarry, Gibraltar
Homo sapiens neanderthalensis,
originally called *Homo calpicus*.

p.79

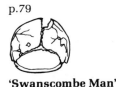

**'Swanscombe Man'
from Swanscombe, England**
Usually called *Homo sapiens
protosapiens*. Also called *Homo
sapiens steinheimensis*.

p.79

From Amud Cave, Israel
Homo sapiens neanderthalensis.

p.80

From La Chapelle-aux-Saints, France
Homo sapiens neanderthalensis,
originally called *Homo chapellensis*.

p.81

From La Ferrassie, France
Homo sapiens neanderthalensis.

Answers

p. 13 The **fallow deer** is a mammal.

p. 15 The **Diana monkey** is a primate.

p. 17 The **gibbon** is an ape.

Index

Acknowledgements

Photographs
37,44: Çandır jaw, Mineral Research and Exploration Institute of Turkey.
50/51: footprints, John Reader, courtesy Dr Mary Leakey and the National Geographical Society.
51: Lucy, Cleveland Museum of Natural History.
54/55,59: car factory, Fiat Auto (UK) Ltd.
71: gas fire, British Gas Corporation.
72: Choukoutien site, Institute of Vertebrate Palaeontology and Palaeo-anthropology, Academia Sinica, Peking.
75: Petralona skull, Dr C. B. Stringer.
76/77,82: funeral of Sir Winston Churchill, Patrick Thurston, Daily Telegraph Colour Library.
82: wedding, D. W. Morbey.
91: Lascaux cave painting, Jean Vertut, courtesy Editions Mazenod (from *Prehistory of Western Art* by Leroi Gourham).
94: feeding sheep, *Farmers Weekly*; pigs, Spectrum Colour Library.
96: Jericho, British School of Archaeology in Jerusalem.
98/99: moon landing composition, NASA photographs AS11–446553, 16–107–17435 and 16–107–17438.

For their cooperation and help in the production of the following photographs, we should like to thank:

25: chromosomes, Dr Joy Delhanty, The Galton Laboratory, Department of Genetics and Biometry, University of London.
26/27: gorilla, Twycross Zoo.
66/67,71: glass-blowing, Whitefriars Glass.
83: christening, Rev R. W. H. Nind, St Matthews, Brixton, London.
88/89,90,91: London Philharmonic Orchestra.
94: vegetables, Covent Garden Market Authority, London.